The Temporary Autonomous Zone, Ontological Anarchy, Poetic Terrorism

Anarchy and Conspiracy

By

Hakim Bey

First published in 1991

Published by Left of Brain Books

Copyright © 2023 Left of Brain Books

ISBN 978-1-397-67001-4

First Edition

PUBLISHER'S PREFACE

About the Book

"The Temporary Autonomous Zone (TAZ) describes the socio-political tactic of creating temporary spaces that elude formal structures of control. The essay uses various historical and philosophical examples, all of which attempt to lead the reader to the conclusion that the best way to create a non-hierarchical system of social relationships is to concentrate on the present and on releasing one's own mind from the controlling mechanisms that have been imposed on it.

In the formation of a TAZ, Bey argues that information becomes a key tool that sneaks into the cracks of formal procedures. A new territory of the moment is created that is on the boundary line of established regions. Any attempt at permanence that goes beyond the moment deteriorates to a structured system that inevitably stifles individual creativity. It is this chance at creativity that is real empowerment."

(Quote from wikipedia.org)

About the Author

Hakim Bey (1945 - Present)

"Peter Lamborn Wilson (b. New York (or Baltimore), 1945) is an American political writer, essayist, and poet, perhaps best known for first proposing the concept of the Temporary Autonomous Zone (TAZ), based on a historical review of pirate utopias. He sometimes writes under the name Hakim Bey. The pseudonym may or may not have been a name-of-convenience or collective pseudonym used by other radical writers since the 1970s and is a combination of the Arabic word for 'wise man'-as well as any "decision-maker" or "ruler"- and a last name common in the Moorish Science Temple. Bey, originally a Turkic word for "chieftain," traditionally applied to the leaders of small tribal groups. In historical accounts, many Turkish, other Turkic and Persian leaders are titled bey, beg or beigh. They are all the same word with the simple meaning of "leader." Also in Turkish, Hakim means judge and Bey is a generic word for a gentleman (mister) generally used after a name."

(Quote from wikipedia.org)

CONTENTS

ACKNOWLEDGMENTS

CHAOS: THE BROADSHEETS OF ONTOLOGICAL ANARCHISM was first published in 1985 by Grim Reaper Press of Weehawken, New Jersey; a later re-issue was published in Providence, Rhode Island, and this edition was pirated in Boulder, Colorado. Another edition was released by Verlag Golem of Providence in 1990, and pirated in Santa Cruz, California, by We Press. "The Temporary Autonomous Zone" was performed at the Jack Kerouac School of Disembodied Poetics in Boulder, and on WBAI-FM in New York City, in 1990.

Thanx to the following publications, current and defunct, in which some of these pieces appeared (no doubt I've lost or forgotten many--sorry!): KAOS (London); Ganymede (London); Pan (Amsterdam); Popular Reality; Exquisite Corpse (also Stiffest of the Corpse, City Lights); Anarchy (Columbia, MO); Factsheet Five; Dharma Combat; OVO; City Lights Review; Rants and Incendiary Tracts (Amok); Apocalypse Culture (Amok); Mondo 2000; The Sporadical; Black Eye; Moorish Science Monitor; FEH!; Fag Rag; The Storm!; Panic (Chicago); Bolo Log (Zurich); Anathema; Seditious Delicious; Minor Problems (London); AQUA; Prakilpana.

Also, thanx to the following individuals: Jim Fleming; James Koehnline; Sue Ann Harkey; Sharon Gannon; Dave Mandl; Bob Black; Robert Anton Wilson; William Burroughs; "P.M."; Joel Birroco; Adam Parfrey; Brett Rutherford; Jake Rabinowitz; Allen Ginsberg; Anne Waldman; Frank Torey; Andr-Codrescu; Dave Crowbar; Ivan Stang; Nathaniel Tarn; Chris Funkhauser; Steve Englander; Alex Trotter. --March, 1991

CHAOS: THE BROADSHEETS OF ONTOLOGI-CAL ANARCHISM

Chaos

CHAOS NEVER DIED. Primordial uncarved block, sole worshipful monster, inert & spontaneous, more ultraviolet than any mythology (like the shadows before Babylon), the original undifferentiated oneness-of-being still radiates serene as the black pennants of Assassins, random & perpetually intoxicated.

Chaos comes before all principles of order & entropy, it's neither a god nor a maggot, its idiotic desires encompass & define every possible choreography, all meaningless aethers & phlogistons: its masks are crystallizations of its own faceless-ness, like clouds.

Everything in nature is perfectly real including consciousness, there's absolutely nothing to worry about. Not only have the chains of the Law been broken, they never existed; demons never guarded the stars, the Empire never got started, Eros never grew a beard.

No, listen, what happened was this: they lied to you, sold you ideas of good & evil, gave you distrust of your body & shame for your prophethood of chaos, invented words of disgust for your molecular love, mesmerized you with inattention, bored you with civilization & all its usurious emotions.

There is no becoming, no revolution, no struggle, no path; already you're the monarch of your own skin--your inviolable freedom waits to be completed only by the love of other monarchs: a politics of dream, urgent as the blueness of sky.

To shed all the illusory rights & hesitations of history demands the economy of some legendary Stone Age--shamans not priests, bards not lords, hunters not police, gatherers of paleolithic laziness, gentle as blood, going naked for a sign or painted as birds, poised on the wave of explicit presence, the clockless nowever.

Agents of chaos cast burning glances at anything or anyone capable of bearing witness to their condition, their fever of lux et voluptas. I am awake only in what I love & desire to the point of terror--everything else is just shrouded furniture, quotidian anaesthesia, shit-for-brains, sub-reptilian ennui of totalitarian regimes, banal censorship & useless pain.

Avatars of chaos act as spies, saboteurs, criminals of amour fou, neither selfless nor selfish, accessible as children, mannered as barbarians, chafed with obsessions, unemployed, sensually deranged, wolfangels, mirrors for contemplation, eyes like flowers, pirates of all signs & meanings.

Here we are crawling the cracks between walls of church state school & factory, all the paranoid monoliths. Cut off from the tribe by feral nostalgia we tunnel after lost words, imaginary bombs.

The last possible deed is that which defines perception itself, an invisible golden cord that connects us: illegal dancing in the courthouse corridors. If I were to kiss you here they'd call it an act of terrorism--so let's take our pistols to bed & wake up the city at midnight like drunken bandits celebrating with a fusillade, the message of the taste of chaos.

Poetic Terrorism

WEIRD DANCING IN ALL-NIGHT computer-banking lobbies. Unauthorized pyrotechnic displays. Land-art, earth-works as bizarre alien artifacts strewn in State Parks. Burglarize houses but instead of stealing, leave Poetic-Terrorist objects. Kidnap someone & make them happy. Pick someone at random & convince them they're the heir to an enormous, useless & amazing fortune--say 5000 square miles of Antarctica, or an aging circus elephant, or an orphanage in Bombay, or a collection of alchemical mss. Later they will come to realize that for a few moments they believed in something extraordinary, & will perhaps be driven as a result to seek out some more intense mode of existence.

Bolt up brass commemorative plaques in places (public or private) where you have experienced a revelation or had a particularly fulfilling sexual experience, etc.

Go naked for a sign.

Organize a strike in your school or workplace on the grounds that it does not satisfy your need for indolence & spiritual beauty.

Grafitti-art loaned some grace to ugly subways & rigid public momuments--PT-art can also be created for public places: poems scrawled in courthouse lavatories, small fetishes abandoned in parks & restaurants, xerox-art under windshield-wipers of parked cars, Big Character Slogans pasted on playground walls, anonymous letters mailed to random or chosen recipients (mail fraud), pirate radio transmissions, wet cement...

The audience reaction or aesthetic-shock produced by PT ought to be at least as strong as the emotion of terror-- powerful disgust, sexual arousal, superstitious awe, sudden intuitive breakthrough, dada-esque angst--no matter whether the PT is aimed at one person or many, no matter whether it is "signed" or anonymous, if it does not change someone's life (aside from the artist) it fails.

PT is an act in a Theater of Cruelty which has no stage, no rows of seats, no tickets & no walls. In order to work at all, PT must categorically be divorced from all conventional structures for art consumption (galleries, publications, media). Even the guerilla Situationist tactics of street theater are perhaps too well known & expected now.

An exquisite seduction carried out not only in the cause of mutual satisfaction but also as a conscious act in a deliberately beautiful life--may be the ultimate PT. The PTerrorist behaves like a confidence-trickster whose aim is not money but CHANGE.

Don't do PT for other artists, do it for people who will not realize (at least for a few moments) that what you have done is art. Avoid recognizable art-categories, avoid politics, don't stick around to argue, don't be sentimental; be ruthless, take risks, vandalize only what must be defaced, do something children will remember all their lives--but don't be spontaneous unless the PT Muse has possessed you.

Dress up. Leave a false name. Be legendary. The best PT is against the law, but don't get caught. Art as crime; crime as art.

Amour Fou

AMOUR FOU IS NOT a Social Democracy, it is not a Parliament of Two. The minutes of its secret meetings deal with meanings too enormous but too precise for prose. Not this, not that--its Book of Emblems trembles in your hand.

Naturally it shits on schoolmasters & police, but it sneers at liberationists & ideologues as well--it is not a clean well-lit room. A topological charlatan laid out its corridors & abandoned parks, its ambush-decor of luminous black & membranous maniacal red.

Each of us owns half the map--like two renaissance potentates we define a new culture with our anathematized mingling of bodies, merging of liquids--the Imaginal seams of our City-state blur in our sweat.

Ontological anarchism never came back from its last fishing trip. So long as no one squeals to the FBI, CHAOS cares nothing for the future of civilization. Amour fou breeds only by accident--its primary goal is ingestion of the Galaxy. A conspiracy of transmutation.

Its only concern for the Family lies in the possibility of incest ("Grow your own!" "Every human a Pharoah!")--O most sincere of readers, my semblance, my brother/sister!--& in the masturbation of a child it finds concealed (like a japanese-paper-flower-pill) the image of the crumbling of the State.

Words belong to those who use them only till someone else steals them back. The Surrealists disgraced themselves by selling amour fou to the ghost-machine of Abstraction--they sought in their unconsciousness only power over others, & in this they followed de Sade (who wanted "freedom" only for grown-up whitemen to eviscerate women & children).

Amour fou is saturated with its own aesthetic, it fills itself to the borders of itself with the trajectories of its own gestures, it runs on angels' clocks, it is not a fit fate for commissars & shopkeepers. Its ego evaporates in the mutability of desire, its communal spirit withers in the selfishness of obsession.

Amour fou involves non-ordinary sexuality the way sorcery demands non-ordinary consciousness. The anglo-saxon post-Protestant world channels all its suppressed sensuality into advertising & splits itself into clashing mobs: hysterical prudes vs promiscuous clones & former-ex-singles. AF doesn't want to join anyone's army, it takes no part in the Gender Wars, it is bored by equal opportunity employment (in fact it refuses to work for a living), it doesn't complain, doesn't explain, never votes & never pays taxes.

AF would like to see every bastard ("lovechild") come to term & birthed--AF thrives on anti-entropic devices--AF loves to be molested by children--AF is better than prayer, better than sinsemilla--AF takes its own palmtrees & moon wherever it goes. AF admires tropicalismo, sabotage, break-dancing, Layla & Majnun, the smells of gunpowder & sperm.

AF is always illegal, whether it's disguised as a marriage or a boyscout troop--always drunk, whether on the wine of its own secretions or the smoke of its own polymorphous virtues. It is not the derangement of the senses but rather their apotheosis--not the result of freedom but rather its precondition. Lux et voluptas.

Wild Children

THE FULL MOON'S UNFATHOMABLE light-path--mid-May midnight in some State that starts with "I," so two-dimensional it can scarcely be said to possess any geography at all--the beams so urgent & tangible you must draw the shades in order to think in words.

No question of writing to Wild Children. They think in images-- prose is for them a code not yet fully digested & ossified, just as for us never fully trusted.

You may write about them, so that others who have lost the silver chain may follow. Or write for them, making of STORY & EMBLEM a process of seduction into your own paleolithic memories, a barbaric enticement to liberty (chaos as CHAOS understands it).

For this otherworld species or "third sex," les enfants sauvages, fancy & Imagination are still undifferentiated. Unbridled PLAY: at one & the same time the source of our Art & of all the race's rarest eros.

To embrace disorder both as wellspring of style & voluptuous storehouse, a fundamental of our alien & occult civilization, our conspiratorial esthetic, our lunatic espionage--this is the action (let's face it) either of an artist of some sort, or of a ten- or thirteen-year-old.

Children whose clarified senses betray them into a brilliant sorcery of beautiful pleasure reflect something feral & smutty in the nature of reality itself: natural ontological anarchists, angels of chaos--their gestures & body odors broadcast around them a jungle of presence, a forest of prescience complete with snakes, ninja weapons, turtles, futuristic shamanism, incredible mess, piss, ghosts, sunlight, jerking off, birds' nests & eggs--gleeful aggression against the groan-ups of those Lower Planes so powerless to englobe either destructive epiphanies or creation in the form of antics fragile but sharp enough to slice moonlight.

And yet the denizens of these inferior jerkwater dimensions truly believe they control the destinies of Wild Children--& down here, such vicious beliefs actually sculpt most of the substance of happenstance.

The only ones who actually wish to share the mischievous destiny of those savage runaways or minor guerillas rather than dictate it, the only ones who can understand that cherishing & unleashing are the same act--these are mostly artists, anarchists, perverts, heretics, a band apart (as much from each other as from the world) or able to meet only as wild children might, locking gazes across a dinnertable while adults gibber from behind their masks.

Too young for Harley choppers--flunk-outs, break-dancers, scarcely pubescent poets of flat lost railroad towns--a million sparks falling from the skyrockets of Rimbaud & Mowgli-- slender terrorists whose gaudy bombs are compacted of polymorphous love & the precious shards of popular culture-- punk gunslingers dreaming of piercing their ears, animist bicyclists gliding in the pewter dusk through Welfare streets of accidental flowers--out-of-season gypsy skinny-dippers, smiling sideways-glancing thieves of power-totems, small change & panther-bladed knives--we sense them everywhere--we publish this offer to trade the corruption of our own lux et gaudium for their perfect gentle filth.

So get this: our realization, our liberation depends on theirs-- not because we ape the Family, those "misers of love" who hold hostages for a banal future, nor the State which schools us all to

sink beneath the event-horizon of a tedious "usefulness"--
no--but because we & they, the wild ones, are images of each
other, linked & bordered by that silver chain which defines the
pale of sensuality, transgression & vision.

We share the same enemies & our means of triumphant escape
are also the same: a delirious & obsessive play, powered by the
spectral brilliance of the wolves & their children.

Paganism

CONSTELLATIONS BY WHICH TO steer the barque of the soul. "If
the moslem understood Islam he would become an idol-
worshipper."--Mahmud Shabestari Eleggua, ugly opener of
doors with a hook in his head & cowrie shells for eyes, black
santeria cigar & glass of rum--same as Ganesh, elephant-head
fat boy of Beginnings who rides a mouse. The organ which
senses the numinous atrophies with the senses. Those who
cannot feel baraka cannot know the caress of the world.

Hermes Poimandres taught the animation of eidolons, the
magic in-dwelling of icons by spirits--but those who cannot
perform this rite on themselves & on the whole palpable fabric
of material being will inherit only blues, rubbish, decay.

The pagan body becomes a Court of Angels who all perceive this
place--this very grove--as paradise ("If there is a paradise,
surely it is here!"--inscription on a Mughal garden gate).

But ontological anarchism is too paleolithic for eschatology--
things are real, sorcery works, bush-spirits one with the
Imagination, death an unpleasant vagueness--the plot of Ovid's
Metamorphoses--an epic of mutability. The personal myth-
scape.

Paganism has not yet invented laws--only virtues. No priestcraft, no theology or metaphysics or morality--but a universal shamanism in which no one attains real humanity without a vision.

Food money sex sleep sun sand & sinsemilla--love truth peace freedom & justice. Beauty. Dionysus the drunk boy on a panther--rank adolescent sweat--Pan goatman slogs through the solid earth up to his waist as if it were the sea, his skin crusted with moss & lichen--Eros multiplies himself into a dozen pastoral naked Iowa farm boys with muddy feet & pond-scum on their thighs.

Raven, the potlatch trickster, sometimes a boy, old woman, bird who stole the Moon, pine needles floating on a pond, Heckle/Jeckle totempole-head, chorus-line of crows with silver eyes dancing on the woodpile--same as Semar the hunchback albino hermaphrodite shadow-puppet patron of the Javanese revolution.

Yemaya, bluestar sea-goddess & patroness of queers--same as Tara, bluegrey aspect of Kali, necklace of skulls, dancing on Shiva's stiff lingam, licking monsoon clouds with her yard-long tongue--same as Loro Kidul, jasper-green Javanese sea-goddess who bestows the power of invulnerability on sultans by tantrik intercourse in magic towers & caves.

From one point of view ontological anarchism is extremely bare, stripped of all qualities & possessions, poor as CHAOS itself--but from another point of view it pullulates with baroqueness like the Fucking-Temples of Kathmandu or an alchemical emblem book--it sprawls on its divan eating loukoum & entertaining heretical notions, one hand inside its baggy trousers.

The hulls of its pirate ships are lacquered black, the lateen sails are red, black banners with the device of a winged hourglass.

A South China Sea of the mind, off a jungle-flat coast of palms, rotten gold temples to unknown bestiary gods, island after island, the breeze like wet yellow silk on naked skin, navigating by pantheistic stars, hierophany on hierophany, light upon light against the luminous & chaotic dark.

Art Sabotage

ART SABOTAGE STRIVES TO be perfectly exemplary but at the same time retain an element of opacity--not propaganda but aesthetic shock--apallingly direct yet also subtly angled-- action-as-metaphor.

Art Sabotage is the dark side of Poetic Terrorism--creation-through-destruction--but it cannot serve any Party, nor any nihilism, nor even art itself. Just as the banishment of illusion enhances awareness, so the demolition of aesthetic blight sweetens the air of the world of discourse, of the Other. Art Sabotage serves only consciousness, attentiveness, awakeness.

A-S goes beyond paranoia, beyond deconstruction--the ultimate criticism--physical attack on offensive art-- aesthetic jihad. The slightest taint of petty ego-icity or even of personal taste spoils its purity & vitiates its force. A-S can never seek power--only release it.

Individual artworks (even the worst) are largely irrelevant--A-S seeks to damage institutions which use art to diminish consciousness & profit by delusion. This or that poet or painter cannot be condemned for lack of vision--but malign Ideas can be assaulted through the artifacts they generate. MUZAK is designed to hypnotize & control--its machinery can be smashed.

Public book burnings--why should rednecks & Customs officials monopolize this weapon? Novels about children possessed by demons; the New York Times bestseller list; feminist tracts against pornography; schoolbooks (especially Social Studies, Civics, Health); piles of New York Post , Village Voice & other supermarket papers; choice gleanings of Xtian publishers; a few Harlequin Romances--a festive atmosphere, wine-bottles & joints passed around on a clear autumn afternoon.

To throw money away at the Stock Exchange was pretty decent Poetic Terrorism--but to destroy the money would have been good Art Sabotage. To seize TV transmission & broadcast a few pirated minutes of incendiary Chaote art would constitute a feat of PT--but simply to blow up the transmission tower would be perfectly adequate Art Sabotage. If certain galleries & museums deserve an occasional brick through their windows--not destruction, but a jolt to complacency--then what about BANKS? Galleries turn beauty into a commodity but banks transmute Imagination into feces and debt. Wouldn't the world gain a degree of beauty with each bank that could be made to tremble...or fall? But how? Art Sabotage should probably stay away from politics (it's so boring)--but not from banks.

Don't picket--vandalize. Don't protest--deface. When ugliness, poor design & stupid waste are forced upon you, turn Luddite, throw your shoe in the works, retaliate. Smash the symbols of the Empire in the name of nothing but the heart's longing for grace.

The Assassins

ACROSS THE LUSTER OF the desert & into the polychrome hills, hairless & ochre violet dun & umber, at the top of a dessicate blue valley travelers find an artificial oasis, a fortified castle in saracenic style enclosing a hidden garden.

As guests of the Old Man of the Mountain Hassan-i Sabbah they climb rock-cut steps to the castle. Here the Day of Resurrection has already come & gone--those within live outside profane Time, which they hold at bay with daggers & poisons.

Behind crenellations & slit-windowed towers scholars & fedayeen wake in narrow monolithic cells. Star-maps, astrolabes, alembics & retorts, piles of open books in a shaft of morning sunlight--an unsheathed scimitar.

Each of those who enter the realm of the Imam-of-one's-own-being becomes a sultan of inverted revelation, a monarch of abrogation & apostasy. In a central chamber scalloped with light and hung with tapestried arabesques they lean on bolsters & smoke long chibouks of haschisch scented with opium & amber.

For them the hierarchy of being has compacted to a dimension-less punctum of the real--for them the chains of Law have been broken--they end their fasting with wine. For them the outside of everything is its inside, its true face shines through direct. But the garden gates are camouflaged with terrorism, mirrors, rumors of assassination, trompe l'oeil, legends.

Pomegranate, mulberry, persimmon, the erotic melancholy of cypresses, membrane-pink shirazi roses, braziers of meccan aloes & benzoin, stiff shafts of ottoman tulips, carpets spread like make-believe gardens on actual lawns--a pavilion set with a mosaic of calligrammes--a willow, a stream with watercress--a fountain crystalled underneath with geometry-- the metaphysical scandal of bathing odalisques, of wet brown cupbearers hide-&-seeking in the foliage--"water, greenery, beautiful faces."

By night Hassan-i Sabbah like a civilized wolf in a turban stretches out on a parapet above the garden & glares at the sky, conning the asterisms of heresy in the mindless cool desert air. True, in this myth some aspirant disciples may be ordered to fling themselves off the ramparts into the black--but also true that some of them will learn to fly like sorcerers.

The emblem of Alamut holds in the mind, a mandals or magic circle lost to history but embedded or imprinted in conscious-ness. The Old Man flits like a ghost into tents of kings & bedrooms of theologians, past all locks & guards with forgotten moslem/ninja techniques, leaves behind bad dreams, stilettos on pillows, puissant bribes.

The attar of his propaganda seeps into the criminal dreams of ontological anarchism, the heraldry of our obsessions displays the luminous black outlaw banners of the Assassins...all of them pretenders to the throne of an Imaginal Egypt, an occult space/light continuum consumed by still-unimagined liberties.

Pyrotechnics

INVENTED BY THE CHINESE but never developed for war--a fine example of Poetic Terrorism--a weapon used to trigger aesthetic shock rather than kill--the Chinese hated war & used to go into mourning when armies were raised--gunpowder more useful to frighten malign demons, delight children, fill the air with brave & risky-smelling haze.

Class C Thunder Bombs from Kwantung, bottlerockets, butterflies, M-80's, sunflowers, "A Forest In Springtime"--revolution weather--light your cigarette from the sizzling fuse of a Haymarket-black bomb--imagine the air full of lamiae &

succubi, oppressive spirits, police-ghosts. Call some kid with a smouldering punk or kitchen match-- shaman-apostle of summer gunpowder plots--shatter the heavy night with pinched stars & pumped stars, arsenic & antimony, sodium & calomel, a blitz of magnesium & shrill picrate of potash.

Spur-fire (lampblack & saltpetre) portfire & iron filings-- attack your local bank or ugly church with roman candles & purple-gold skyrockets, impromptu & anonymous (perhaps launch from back of pick-up truck..)

Build frame-lattice lancework set-pieces on the roofs of insurance buildings or schools--a kundalini-snake or Chaos-dragon coiled barium-green against a background of sodium-oxalate yellow--Don't Tread On Me--or copulating monsters shooting wads of jizm-fire at a Baptists old folks home.

Cloud-sculpture, smoke sculpture & flags = Air Art. Earthworks. Fountains = Water Art. And Fireworks. Don't perform with Rockefeller grants & police permits for audiences of culture-lovers. Evanescent incendiary mind-bombs, scary mandalas flaring up on smug suburban nights, alien green thunderheads of emotional plague blasted by orgone-blue vajra-rays of lasered feux d'artifice.

Comets that explode with the odor of hashish & radioactive charcoal--swampghouls & will-o'-the-wisps haunting public parks--fake St. Elmo's fire flickering over the architecture of the bourgeoisie--strings of lady-fingers falling on the Legislature floor--salamander-elementals attack well-known moral reformers.

Blazing shellac, sugar of milk, strontium, pitch, gum water, gerbs of chinese fire--for a few moments the air is ozone-sharp--drifting opal cloud of pungent dragon/phoenix smoke. For an instant the Empire falls, its princes & governors flee to their stygian muck, plumes of sulphur from elf-flamethrowers burning their pinched asses as they retreat. The Assassin-child, psyche of fire, holds sway for one brief dogstar-hot night.

Chaos Myths

Unseen Chaos (po-te-kitea)
Unpossessed, Unpassing
Chaos of utter darkness
Untouched & untouchable
--Maori Chant

Chaos perches on a sky-mountain: a huge bird like a yellow bag or red fireball, with six feet & four wings--has no face but dances & sings.

Or Chaos is a black longhaired dog, blind & deaf, lacking the five viscera.

Chaos the Abyss comes first, then Earth/Gaia, then Desire/Eros. From these three proceed two pairs--Erebus & old Night, Aether & Daylight. Neither Being nor Non-being
neither air nor earth nor space:
what was enclosed? where? under whose protection?
What was water, deep, unfathomable?
Neither death nor immortality, day nor night--
but ONE breathed by itself with no wind.
Nothing else. Darkness swathed in darkness,
unmanifest water.
The ONE, hidden by void,
felt the generation of heat, came into being
as Desire, first seed of Mind...
Was there an up or down?

There were casters of seed, there were powers:
energy underneath, impulse above.
But who knows for sure?
--Rg Veda

Tiamat the Chaos-Ocean slowly drops from her womb Silt & Slime, the Horizons, Sky and watery Wisdom. These offspring grow noisy & bumptious--she considers their destruction.

But Marduk the wargod of Babylon rises in rebellion against the Old Hag & her Chaos-monsters, chthonic totems--Worm, Female Ogre, Great Lion, Mad Dog, Scorpion Man, Howling Storm--dragons wearing their glory like gods--& Tiamat herself a great sea-serpent.

Marduk accuses her of causing sons to rebel against fathers-- she loves Mist & Cloud, principles of disorder. Marduk will be the first to rule, to invent government. In battle he slays Tiamat & from her body orders the material universe. He inaugurates the Babylonian Empire--then from gibbets & bloody entrails of Tiamat's incestuous son he creates the human race to serve forever the comfort of gods--& their high priests & anointed kings.

Father Zeus & the Olympians wage war against Mother Gaia & the Titans, those partisans of Chaos, the old ways of hunting & gathering, of aimless wandering, androgyny & the license of beasts.

Amon-Ra (Being) sits alone in the primordial Chaos-Ocean of NUN creating all the other gods by jerking off--but Chaos also manifests as the dragon Apophis whom Ra must destroy (along with his state of glory, his shadow & his magic) in order that the Pharoah may safely rule--a victory ritually re-created daily in

Imperial temples to confound the enemies of the State, of cosmic Order.

Chaos is Hun Tun, Emperor of the Center. One day the South Sea, Emperor Shu, & the North Sea, Emperor Hu (shu hu = lightning) paid a visit to Hun Tun, who always treated them well. Wishing to repay his kindness they said, "All beings have seven orifices for seeing, hearing, eating, shitting, etc.--but poor old Hun Tun has none! Let's drill some into him!" So they did--one orifice a day--till on the seventh day, Chaos died.

But...Chaos is also an enormous chicken's egg. Inside it P'an-Ku is born & grows for 18,000 years--at last the egg opens up, splits into sky & earth, yang & yin. Now P'an-Ku grows into a column that holds up the universe--or else he becomes the universe (breath-->wind, eyes-->sun & moon, blood & humors-->rivers & seas, hair & lashes-->stars & planets, sperm-->pearls, marrow-->jade, his fleas-->human beings, etc.)

Or else he becomes the man/monster Yellow Emperor. Or else he becomes Lao Tzu, prophet of Tao. In fact, poor old Hun Tun is the Tao itself.

"Nature's music has no existence outside things. The various apertures, pipes, flutes, all living beings together make up nature. The "I" cannot produce things & things cannot produce the "I," which is self-existent. Things are what they are spontaneously, not caused by something else. Everything is natural & does not know why it is so. The 10,000 things have 10,000 different states, all in motion as if there were a True Lord to move them--but if we search for evidence of this Lord we fail to find any." (Kuo Hsiang)

Every realized consciousness is an "emperor" whose sole form of rule is to do nothing to disturb the spontaneity of nature, the Tao. The "sage" is not Chaos itself, but rather a loyal child of Chaos--one of P'an-Ku's fleas, a fragment of flesh of Tiamat's monstrous son. "Heaven and Earth," says Chuang Tzu, "were born at the same time I was, & the 10,000 things are one with me."

Ontological Anarchism tends to disagree only with the Taoists' total quietism. In our world Chaos has been overthrown by younger gods, moralists, phallocrats, banker-priests, fit lords for serfs. If rebellion proves impossible then at least a kind of clandestine spiritual jihad might be launched. Let it follow the war-banners of the anarchist black dragon, Tiamat, Hun Tun.

Chaos never died.

Pornography

IN PERSIA I SAW that poetry is meant to be set to music & chanted or sung--for one reason alone--because it works.

A right combination of image & tune plunges the audience into a hal (something between emotional/aesthetic mood & trance of hyperawareness), outbursts of weeping, fits of dancing--measurable physical response to art. For us the link between poetry & body died with the bardic era--we read under the influence of a cartesian anaesthetic gas.

In N. India even non-musical recitation provokes noise & motion, each good couplet applauded, "Wa! Wa!" with elegant hand-jive, tossing of rupees--whereas we listen to poetry like some SciFi brain in a jar--at best a wry chuckle or grimace, vestige of simian rictus--the rest of the body off on some other planet.

In the East poets are sometimes thrown in prison--a sort of compliment, since it suggests the author has done something at least as real as theft or rape or revolution. Here poets are allowed to publish anything at all--a sort of punishment in effect, prison without walls, without echoes, without palpable existence--shadow-realm of print, or of abstract thought--world without risk or eros.

So poetry is dead again--& even if the mumia from its corpse retains some healing properties, auto-resurrection isn't one of them.

If rulers refuse to consider poems as crimes, then someone must commit crimes that serve the function of poetry, or texts that possess the resonance of terrorism. At any cost re-connect poetry to the body. Not crimes against bodies, but against Ideas (& Ideas-in-things) which are deadly & suffocating. Not stupid libertinage but exemplary crimes, aesthetic crimes, crimes for love. In England some pornographic books are still banned. Pornography has a measurable physical effect on its readers. Like propaganda it sometimes changes lives because it uncovers true desires.

Our culture produces most of its porn out of body-hatred-- but erotic art in itself makes a better vehicle for enhancement of bing/consciousness/bliss--as in certain oriental works. A sort of Western tantrik porn might help galvanize the corpse, make it shine with some of the glamor of crime.

America has freedom of speech because all words are consi-dered equally vapid. Only images count--the censors love snaps of death & mutilation but recoil in horror at the sight of a child masturbating--apparently they experience this as an invasion of their existential validity, their identification with the Empire & its subtlest gestures.

No doubt even the most poetic porn would never revive the faceless corpse to dance & sing (like the Chinese Chaos-bird)--but...imagine a script for a three-minute film set on a mythical isle of runaway children who inhabit ruins of old castles or build totem-huts & junk-assemblage nests--mixture of animation, special-effects, compugraphix & color tape--edited tight as a fastfood commercial...

...but weird & naked, feathers & bones, tents sewn with crystal, black dogs, pigeon-blood--flashes of amber limbs tangled in sheets--faces in starry masks kissing soft creases of skin--androgynous pirates, castaway faces of columbines sleeping on thigh-white flowers--nasty hilarious piss jokes, pet lizards lapping spilt milk--nude break-dancing--victorian bathtub with rubber ducks & pink boners-- Alice on ganja...

...atonal punk reggae scored for gamelan, synthesizer, saxophones & drums--electric boogie lyrics sung by aetherial children's choir--ontological anarchist lyrics, cross between Hafez & Pancho Villa, Li Po & Bakunin, Kabir & Tzara--call it "CHAOS--the Rock Video!"

No...probably just a dream. Too expensive to produce, & besides, who would see it? Not the kids it was meant to seduce. Pirate TV is a futile fantasy, rock merely another commodity--forget the slick gesamtkunstwerk, then. Leaflet a playground with inflammatory smutty feuilletons--pornopropaganda, crackpot samizdat to unchain Desire from its bondage.

Crime

JUSTICE CANNOT BE OBTAINED under any Law--action in accord with spontaneous nature, action which is just, cannot be defined by dogma. The crimes advocated in these broadsheets cannot be committed against self or other but only against the mordant crystallization of Ideas into structures of poisonous Thrones & Dominations.

That is, not crimes against nature or humanity but crimes by legal fiat. Sooner or later the uncovering & unveiling of self/nature transmogrifies a person into a brigand--like stepping into another world then returning to this one to discover you've been declared a traitor, heretic, exile. The Law waits for you to stumble on a mode of being, a soul different from the FDA-approved purple-stamped standard dead meat--& as soon as you begin to act in harmony with nature the Law garottes & strangles you--so don't play the blessed liberal middleclass martyr--accept the fact that you're a criminal & be prepared to act like one.

Paradox: to embrace Chaos is not to slide toward entropy but to emerge into an energy like stars, a pattern of instantaneous grace--a spontaneous organic order completely different from the carrion pyramids of sultans, muftis, cadis & grinning executioners.

After Chaos comes Eros--the principle of order implicit in the nothingness of the unqualified One. Love is structure, system, the only code untainted by slavery & drugged sleep. We must become crooks & con-men to protect its spiritual beauty in a bezel of clandestinity, a hidden garden of espionage.

Don't just survive while waiting for someone's revolution to clear your head, don't sign up for the armies of anorexia or bulimia--act as if you were already free, calculate the odds, step out, remember the Code Duello--Smoke Pot/Eat Chicken/Drink Tea. Every man his own vine & figtree (Circle Seven Koran, Noble Drew Ali)--carry your Moorish passport with pride, don't get caught in the crossfire, keep your back covered--but take the risk, dance before you calcify.

The natural social model for ontological anarchism is the child-gang or the bank-robbers-band. Money is a lie--this adventure must be feasible without it--booty & pillage should be spent before it turns back into dust. Today is Resurrection Day--money wasted on beauty will be alchemically transmuted into elixir. As my uncle Melvin used to say, stolen watermelon tastes sweeter. The world is already re-made according to the heart's desire--but civilization owns all the leases & most of the guns. Our feral angels demand we trespass, for they manifest themselves only on forbidden grounds. High Way Man. The yoga of stealth, the lightning raid, the enjoyment of treasure.

Sorcery

THE UNIVERSE WANTS TO PLAY. Those who refuse out of dry spiritual greed & choose pure contemplation forfeit their humanity--those who refuse out of dull anguish, those who hesitate, lose their chance at divinity--those who mold themselves blind masks of Ideas & thrash around seeking some proof of their own solidity end by seeing out of dead men's eyes.

Sorcery: the systematic cultivation of enhanced consciousness or non-ordinary awareness & its deployment in the world of deeds & objects to bring about desired results.

The incremental openings of perception gradually banish the false selves, our cacophonous ghosts--the "black magic" of envy & vendetta backfires because Desire cannot be forced. Where our knowledge of beauty harmonizes with the ludus naturae, sorcery begins.

No, not spoon-bending or horoscopy, not the Golden Dawn or make-believe shamanism, astral projection or the Satanic Mass--if it's mumbo jumbo you want go for the real stuff, banking, politics, social science--not that weak blavatskian crap.

Sorcery works at creating around itself a psychic/physical space or openings into a space of untrammeled expression--the metamorphosis of quotidian place into angelic sphere. This involves the manipulation of symbols (which are also things) & of people (who are also symbolic)--the archetypes supply a vocabulary for this process & therefore are treated as if they were both real & unreal, like words. Imaginal Yoga.

The sorcerer is a Simple Realist: the world is real--but then so must consciousness be real since its effects are so tangible. The dullard finds even wine tasteless but the sorcerer can be intoxicated by the mere sight of water. Quality of perception defines the world of intoxication--but to sustain it & expand it to include others demands activity of a certain kind--sorcery. Sorcery breaks no law of nature because there is no Natural Law, only the spontaneity of natura naturans, the tao. Sorcery violates laws which seek to chain this flow--priests, kings, hierophants, mystics, scientists & shopkeepers all brand the sorcerer enemy for threatening the power of their charade, the tensile strength of their illusory web.

A poem can act as a spell & vice versa--but sorcery refuses to be a metaphor for mere literature--it insists that symbols must cause events as well as private epiphanies. It is not a critique but a re-making. It rejects all eschatology & metaphysics of removal, all bleary nostalgia & strident futurismo, in favor of a paroxysm or seizure of presence.

Incense & crystal, dagger & sword, wand, robes, rum, cigars, candles, herbs like dried dreams--the virgin boy staring into a bowl of ink--wine & ganja, meat, yantras & gestures--rituals of pleasure, the garden of houris & sakis--the sorcerer climbs these snakes & ladders to a moment which is fully saturated with its own color, where mountains are mountains & trees are trees, where the body becomes all time, the beloved all space.

The tactics of ontological anarchism are rooted in this secret Art--the goals of ontological anarchism appear in its flowering. Chaos hexes its enemies & rewards its devotees...this strange yellowing pamphlet, pseudonymous & dust-stained, reveals all...send away for one split second of eternity.

Advertisement

WHAT THIS TELLS YOU is not prose. It may be pinned to the board but it's still alive & wriggling. It does not want to seduce you unless you're extremely young & good-looking (enclose recent photo).

Hakim Bey lives in a seedy Chinese hotel where the proprietor nods out over newspaper & scratchy broadcasts of Peking Opera. The ceiling fan turns like a sluggish dervish--sweat falls on the page--the poet's kaftan is rusty, his ovals spill ash on the rug--his monologues seem disjointed & slightly sinister--outside shuttered windows the barrio fades into palmtrees, the naive blue ocean, the philosophy of tropicalismo.

Along a highway somewhere east of Baltimore you pass an Airstream trailer with a big sign on the lawn SPIRITUAL READINGS & the image of a crude black hand on a red background. Inside you notice a display of dream-books, numbers-books, pamphlets on HooDoo and Santeria, dusty old nudist magazines, a pile of Boy's Life, treatises on fighting-cocks...& this book, Chaos. Like words spoken in a dream, portentous, evanescent, changing into perfumes, birds, colors, forgotten music.

This book distances itself by a certain impassibility of surface, almost a glassiness. It doesn't wag its tail & it doesn't snarl but it bites & humps the furniture. It doesn't have an ISBN number & it doesn't want you for a disciple but it might kidnap your children.

This book is nervous like coffee or malaria--it sets up a network of cut-outs & safe drops between itself & its readers--but it's so baldfaced & literal-minded it practically encodes itself--it smokes itself into a stupor.

A mask, an automythology, a map without placenames--stiff as an egyptian wallpainting nevertheless it reaches to caress someone's face--& suddenly finds itself out in the street, in a body, embodied in light, walking, awake, almost satisfied.

--NYC, May 1-July 4, 1984

COMMUNIQUES OF THE ASSOCIATION FOR ONTOLOGICAL ANARCHY

COMMUNIQUE #1 (SPRING 1986)

I. Slogans & Mottos for Subway Graffiti & Other Purposes

ROOTLESS COSMOPOLITANISM
POETIC TERRORISM
(for scrawling or rubberstamping on advertisements:)
THIS IS YOUR TRUE DESIRE
MARXISM-STIRNERISM
STRIKE FOR INDOLENCE & SPIRITUAL BEAUTY
YOUNG CHILDREN HAVE BEAUTIFUL FEET
THE CHAINS OF LAW HAVE BEEN BROKEN
TANTRIK PORNOGRAPHY
RADICAL ARISTOCRATISM
KIDS' LIB URBAN GUERILLAS
IMAGINARY SHIITE FANATICS
BOLO'BOLO
GAY ZIONISM
(SODOM FOR THE SODOMITES)
PIRATE UTOPIAS
CHAOS NEVER DIED

Some of these are "sincere" slogans of the A.O.A.--others are meant to rouse public apprehension & misgivings--but we're not sure which is which. Thanx to Stalin, Anon., Bob Black, Pir Hassan (upon his mention be peace), F. Nietzsche, Hank Purcell Jr., "P.M.," & Bro. Abu Jehad al-Salah of the Moorish Temple of Dagon.

II. Some Poetic-Terrorist Ideas Still Sadly Languishing in the Realm of "Conceptual Art"

1. Walk into Citibank or Chembank computer customer service area during busy period, take a shit on the floor, & leave.

2. Chicago May Day '86: organize "religious" procession for Haymarket "Martyrs"--huge banners with sentimental portraits, wreathed in flowers & streaming with tinsel & ribbon, borne by penitenti in black KKKatholic-style hooded gowns--outrageous campy TV acolytes with incense & holy water sprinkle the crowd--anarchists w/ash-smeared faces beat themselves with little flails & whips--a "Pope" in black robes blesses tiny symbolic coffins reverently carried to Cemetery by weeping punks. Such a spectacle ought to offend nearly everyone.

3. Paste up in public places a xerox flyer, photo of a beautiful twelve-year-old boy, naked and masturbating, clearly titled: THE FACE OF GOD.

4. Mail elaborate & exquisite magickal "blessings" anonymously to people or groups you admire, e.g. for their politics or spirituality or physical beauty or success in crime, etc. Follow the same general procedure as outlined in Section 5 below, but utilize an aesthetic of good fortune, bliss or love, as appropriate.

5. Invoke a terrible curse on a malign institution, such as the New York Post or the MUZAK company. A technique adapted from Malaysian sorcerers: send the Company a package containing a bottle, corked and sealed with black wax. Inside: dead insects, scorpions, lizards or the like; a bag containing graveyard dirt ("gris-gris" in American HooDoo terminology) along with other noxious substances; an egg, pierced with iron

nails and pins; and a scroll on which an emblem is drawn (see pp. 59-60).

(This yantra or veve invokes the Black Djinn, the Self's dark shadow. Full details obtainable from the A.O.A.) An accompanying note explains that the hex is sent against the institution & not against individuals--but unless the institution itself ceases to be malign, the curse (like a mirror) will begin to infect the premises with noxious fortune, a miasma of negativity. Prepare a "news release" explaining the curse & taking credit for it in the name of the American Poetry Society. Mail copies of this text to all employees of the institution & to selected media. The night before these letters arrive, wheatpaste the institutional premises with xerox copies of the Black Djinn's emblem, where they will be seen by all employees arriving for work next morning.

(Thanx to Abu Jehad again, & to Sri Anamananda--the Moorish Castellan of Belvedere Weather Tower--& other comrades of the Central Park Autonomous Zone, & Brooklyn Temple Number 1)

COMMUNIQUE #2

The Kallikak Memorial Bolo & Chaos Ashram: A Proposal

NURSING AN OBSESSION FOR Airstream trailers--those classic miniature dirigibles on wheels--& also the New Jersey Pine Barrens, huge lost backlands of sandy creeks & tar pines, cranberry bogs & ghost towns, population around 14 per sq. mile, dirt roads overgrown with fern, brokenspine cabins & isolated rusty mobile homes with burnt-out cars in the front yards land of the mythical Kallikaks--Piney families studied by eugenicists in the 1920's to justify sterilization of rural poor. Some Kallikaks married well, prospered, & waxed bourgeois thanx to good genes--others however never worked real jobs but lived off the woods--incest, sodomy, mental deficiencies galore--photos touched up to make them look vacant & morose--descended from rogue Indians, Hessian mercenaries, rum smugglers, deserters--Lovecraftian degenerates

come to think of it the Kallikaks might well have produced secret Chaotes, precursor sex radicals, Zerowork prophets. Like other monotone landscapes (desert, sea, swamp), the Barrens seem infused with erotic power--not vril or orgone so much as a languid disorder, almost a sluttishness of Nature, as if the very ground & water were formed of sexual flesh, membranes, spongy erectile tissue. We want to squat there, maybe an abandoned hunting/fishing lodge with old woodstove & privy-- or decaying Vacation Cabins on some disused County Highway-- or just a woodlot where we park 2 or 3 Airstreams hidden back in the pines near creek or swimming hole. Were the Kallikaks onto something good? We'll find out

somewhere boys dream that extraterrestrials will come to rescue them from their families, perhaps vaporizing the parents with some alien ray in the process. Oh well. Space Pirate Kidnap Plot Uncovered--"Alien" Unmasked As Shiite Fanatic Queer Poet--UFOs Seen Over Pine Barrens--"Lost Boys Will Leave Earth," Claims So-Called Prophet Of Chaos Hakim Bey

runaway boys, mess & disorder, ecstasy & sloth, skinny-dipping, childhood as permanent insurrection--collections of frogs, snails, leaves--pissing in the moonlight--11, 12, 13--old enough to seize back control of one's own history from parents, school, Welfare, TV--Come live with us in the Barrens--we'll cultivate a local brand of seedless rope to finance our luxuries & contemplation of summer's alchemy--& otherwise produce nothing but artifacts of Poetic Terrorism & mementos of our pleasures

going for aimless rides in the old pickup, fishing & gathering, lying around in the shade reading comics & eating grapes--this is our economy. The suchness of things when unchained from the Law, each molecule an orchid, each atom a pearl to the attentive consciousness--this is our cult. The Airstream is draped with Persian rugs, the lawn is profuse with satisfied weeds

the treehouse becomes a wooden spaceship in the nakedness of July & midnight, half-open to the stars, warm with epicurean sweat, rushed & then hushed by the breathing of the Pines. (Dear Bolo Log: You asked for a practical & feasible utopia--here it is, no mere post-holocaust fantasy, no castles on the moon of Jupiter--a scheme we could start up tomorrow--except that every single aspect of it breaks some law, reveals some absolute taboo in U.S. society, threatens the very fabric of etc., etc. Too bad. This is our true desire, & to attain it we must contemplate not only a life of pure art but also pure crime, pure insurrection. Amen.)

(Thanx to the Grim Reaper & other members of the Si Fan Temple of Providence for YALU, GANO, SILA, & ideas)

COMMUNIQUE #3

Haymarket Issue

"I NEED ONLY MENTION in passing that there is a curious reappearance of the Catfish tradition in the popular Godzilla cycle of films which arose after the nuclear chaos unleashed upon Japan. In fact, the symbolic details in the evolution of

Godzilla filmic poplore parallel in a quite surprising way the traditional Japanese and Chinese mythological and folkloric themes of combat with an ambivalent chaos creature (some of the films, like Mothra, directly recalling the ancient motifs of the cosmic egg/gourd/cocoon) that is usually tamed, after the failure of the civilizational order, through the special and indirect agency of children."--Girardot, Myth & Meaning in Early Taoism: The Theme of Chaos (hun- t'un)

In some old Moorish Science Temple (in Chicago or Baltimore) a friend claimed to have seen a secret altar on which rested a matched pair of six shooters (in velvet-lined case) & a black fez. Supposedly initiation to the inner circle required the neophyte Moor to assassinate at least one cop. /// What about Louis Lingg? Was he a precursor of Ontological Anarchism? "I despise you"--one can't help but admiring such sentiments. But the man dynamited himself aged 22 to cheat the gallows...this is not exactly our chosen path. /// The IDEA of the POLICE like hydra grows 100 new heads for each one cut off--and all these heads are live cops. Slicing off heads gains us nothing, but only enhances the beast's power till it swallows us. /// First murder the IDEA--blow up the monument inside us--& then per-haps...the balance of power will shift. When the last cop in our brain is gunned down by the last unfulfilled desire-- perhaps even the landscape around us will begin to change.../// Poetic Terrorism proposes this sabotage of archetypes as the only practical insurrectionary tactic for the present. But as Shiite Extremists eager for the overthrow (by any means) of all police, ayatollahs, bankers, executioners, priests, etc., we reserve the option of venerating even the "failures" of radical excess. /// A few days unchained from the Empire of Lies might well be worth considerable sacrifice; a moment of exalted realization may outweigh a lifetime of microcephalic boredom & work. /// But this moment must become ours--and our ownership of it is seriously compromised if we must commit suicide to preserve its integrity. So we mix our veneration with irony--it's not martyrdom itself we propose, but the courage of the dynamiter, the self-possession of a Chaos-monster, the attainment of criminal & illegal pleasures.

COMMUNIQUE #4

The End of the World

THE A.O.A. DECLARES ITSELF officially bored with the End of the World. The canonical version has been used since 1945 to keep us cowering in fear of Mutual Assured Destruction & in snivelling servitude to our super-hero politicians (the only ones capable of handling deadly Green Kryptonite)...

What does it mean that we have invented a way to destroy all life on Earth? Nothing much. We have dreamed this as an escape from the contemplation of our own individual deaths. We have made an emblem to serve as the mirror-image of a discarded immortality. Like demented dictators we swoon at the thought of taking it all down with us into the Abyss.

The unofficial version of the Apocalypse involves a lascivious yearning for the End, & for a post-Holocaust Eden where the Survivalists (or the 144,000 Elect of Revelations) can indulge themselves in orgies of Dualist hysteria, endless final confrontations with a seductive evil...

We have seen the ghost of Rene Guenon, cadaverous & topped with a fez (like Boris Karloff as Ardis Bey in The Mummy) leading a funereal No Wave Industrial-Noise rock band in loud buzzing blackfly-chants for the death of Culture & Cosmos: the elitist fetishism of pathetic nihilists, the Gnostic self-disgust of "post-sexual" intellectoids.

Are these dreary ballads not simply mirror-images of all those lies & platitudes about Progress & the Future, beamed from every loudspeaker, zapped like paranoid brain-waves from every schoolbook & TV in the world of the Consensus? The thanatosis of the Hip Millenarians extrudes itself like pus from the false health of the Consumers' & Workers' Paradises.

Anyone who can read history with both hemispheres of the brain knows that a world comes to an end every instant--the waves of time leave washed up behind themselves only dry memories of a closed & petrified past--imperfect memory, itself already dying & autumnal. And every instant also gives birth to a world--despite the cavillings of philosophers & scientists whose bodies have grown numb--a present in which all impossibilities are renewed, where regret & premonition fade to nothing in one presential hologrammatical psychomantric gesture.

The "normative" past or the future heat-death of the universe mean as little to us as last year's GNP or the withering away of the State. All Ideal pasts, all futures which have not yet come to pass, simply obstruct our consciousness of total vivid presence.

Certain sects believe that the world (or "a" world) has already come to an end. For Jehovah's Witnesses it happened in 1914 (yes folks, we are living in the Book of Revelations now). For certain oriental occultists, it occurred during the Major Conjunction of the Planets in 1962. Joachim of Fiore proclaimed the Third Age, that of the Holy Spirit, which replaced those of Father & Son. Hassan II of Alamut proclaimed the Great Resurrection, the immanentization of the eschaton, paradise on earth. Profane time came to an end somewhere in the late

Middle Ages. Since then we've been living angelic time--only most of us don't know it.

Or to take an even more Radical Monist stance: Time never started at all. Chaos never died. The Empire was never founded. We are not now & never have been slaves to the past or hostages to the future.

We suggest that the End of the World be declared a fait accompli; the exact date is unimportant. The ranters in 1650 knew that the Millenium comes now into each soul that wakes to itself, to its own centrality & divinity. "Rejoice, fellow creature," was their greeting. "All is ours!"

I want no part of any other End of the World. A boy smiles at me in the street. A black crow sits in a pink magnolia tree, cawing as orgone accumulates & discharges in a split second over the city...summer begins. I may be your lover...but I spit on your Millenium.

COMMUNIQUE #5

"Intellectual S/M Is the Fascism of the Eighties--The Avant-Garde Eats Shit and Likes It"

COMRADES!

Recently some confusion about "Chaos" has plagued the A.O.A. from certain revanchist quarters, forcing us (who despise polemics) at last to indulge in a Plenary Session devoted to denunciations ex cathedra, portentous as hell; our faces burn red with rhetoric, spit flies from our lips, neck veins bulge with pulpit fervor. We must at last descend to flying banners with angry slogans (in 1930's type faces) declaring what Ontological Anarchy is not.

Remember, only in Classical Physics does Chaos have anything to do with entropy, heat-death, or decay. In our physics (Chaos Theory), Chaos identifies with tao, beyond both yin-as-entropy & yang-as-energy, more a principle of continual creation than of any nihil, void in the sense of potentia, not exhaustion. (Chaos as the "sum of all orders.")

From this alchemy we quintessentialize an aesthetic theory. Chaote art may act terrifying, it may even act grand guignol, but it can never allow itself to be drenched in putrid negativity, thanatosis, schadenfreude (delight in the misery of others), crooning over Nazi memorabilia & serial murders. Ontological Anarchy collects no snuff films & is bored to tears with dominatrices who spout french philosophy. ("Everything is hopeless & I knew it before you did, asshole. Nyahh!")

Wilhelm Reich was driven half mad & killed by agents of the Emotional Plague; maybe half his work derived from sheer paranoia (UFO conspiracies, homophobia, even his orgasm theory), BUT on one point we agree wholeheartedly--sexpol: sexual repression breeds death obsession, which leads to bad politics. A great deal of avant-garde Art is saturated with Deadly Orgone Rays (DOR). Ontological Anarchy aims to build aesthetic cloud-busters (OR-guns) to disperse the miasma of cerebral sado-masochism which now passes for slick, hip, new, fashionable. Self-mutilating "performance" artists strike us as banal & stupid--their art makes everyone more unhappy. What kind of two-bit conniving horseshit...what kind of cockroach-brained Art creeps cooked up this apocalypse stew?

Of course the avant-garde seems "smart"--so did Marinetti & the Futurists, so did Pound & Celine. Compared to that kind of intelligence we'd choose real stupidity, bucolic New Age blissed-out inanity--we'd rather be pinheads than queer for death. But luckily we don't have to scoop out our brains to attain our own queer brand of satori. All the faculties, all the senses belong to us as our property--both heart & head, intellect & spirit, body & soul. Ours is no art of mutilation but of excess, superabundance, amazement.

The purveyors of pointless gloom are the Death Squads of contemporary aesthetics--& we are the "disappeared ones." Their make-believe ballroom of occult 3rd-Reich bric-a-brac & child murder attracts the manipulators of the Spectacle--death looks better on TV than life--& we Chaotes, who preach an insurrectionary joy, are edged out towards silence.

Needless to say we reject all censorship by Church & State--but "after the revolution" we would be willing to take individual & personal responsibility for burning all the Death Squad snuff-art crap & running them out of town on a rail. (Criticism becomes direct action in an anarchist context.) My space has room neither for Jesus & his lords of the flies nor for Chas. Manson & his literary admirers. I want no mundane police--I want no cosmic axe-murderers either; no TV chainsaw massacres, no sensitive poststructuralist novels about necrophilia.

As it happens, the A.O.A. can scarcely hope to sabotage the suffocating mechanisms of the State & its ghostly circuitry--but we just might happen to find ourselves in a position to do something about lesser manifestations of the DOR plague such as the Corpse-Eaters of the Lower East Side & other Art scum. We support artists who use terrifying material in some "higher cause"--who use loving/sexual material of any kind, however shocking or illegal--who use their anger & disgust & their true desires to lurch toward self-realization & beauty & adventure. "Social Nihilism," yes--but not the dead nihilism of gnostic self-disgust. Even if it's violent & abrasive, anyone with a vestigial

3rd eye can see the differences between revolutionary pro-life art & reactionary pro-death art. DOR stinks, & the chaote nose can sniff it out--just as it knows the perfume of spiritual/sexual joy, however buried or masked by other darker scents. Even the Radical Right, for all its horror of flesh & the senses, occasionally comes up with a moment of perception & consciousness-enhancement--but the Death Squads, for all their tired lip service to fashionable revolutionary abstractions, offer us about as much true libertarian energy as the FBI, FDA, or the double-dip Baptists.

We live in a society which advertises its costliest commodities with images of death & mutilation, beaming them direct to the reptilian back-brain of the millions thru alpha-wave-generating carcinogenic reality-warping devices--while certain images of life (such as our favorite, a child masturbating) are banned & punished with incredible ferocity. It takes no guts at all to be an Art Sadist, for salacious death lies at the aesthetic center of our Consensus Paradigm. "Leftists" who like to dress up & play Police-&-Victim, people who jerk off to atrocity photos, people who like to think & intellectualize about splatter art & highfalu-tin hopelessness & groovy ghoulishness & other people's misery--such "artists" are nothing but police-without-power (a perfect definition for many "revolutionaries" too). We have a black bomb for these aesthetic fascists--it explodes with sperm & firecrackers, raucous weeds & piracy, weird Shiite heresies & bubbling paradise-fountains, complex rhythms, pulsations of life, all shapeless & exquisite.

Wake up! Breathe! Feel the world's breath against your skin! Seize the day! Breathe! Breathe!

(Thanx to J. Mander's Four Arguments for the Abolition of Television; Adam Exit; & the Moorish Cosmopolitan of Williamsburg)

COMMUNIQUE #6

I. Salon Apocalypse: "Secret Theater"

AS LONG AS NO Stalin breathes down our necks, why not make some art in the service of...an insurrection?

Never mind if it's "impossible." What else can we hope to attain but the "impossible"? Should we wait for someone else to reveal our true desires?

If art has died, or the audience has withered away, then we find ourselves free of two dead weights. Potentially, everyone is now some kind of artist--& potentially every audience has regained its innocence, its ability to become the art that it experiences.

Provided we can escape from the museums we carry around inside us, provided we can stop selling ourselves tickets to the galleries in our own skulls, we can begin to contemplate an art which re-creates the goal of the sorcerer: changing the structure of reality by the manipulation of living symbols (in this case, the images we've been "given" by the organizers of this salon--murder, war, famine, & greed).

We might now contemplate aesthetic actions which possess some of the resonance of terrorism (or "cruelty," as Artaud put it) aimed at the destruction of abstractions rather than people, at liberation rather than power, pleasure rather than profit, joy rather than fear. "Poetic Terrorism." Our chosen images have the potency of darkness--but all images are masks, & behind these masks lie energies we can turn toward light & pleasure.

For example, the man who invented aikido was a samurai who became a pacifist & refused to fight for Japanese imperialism. He became a hermit, lived on a mountain sitting under a tree.

One day a former fellow-officer came to visit him & accused him of betrayal, cowardice, etc. The hermit said nothing, but kept on sitting--& the officer fell into a rage, drew his sword, & struck. Spontaneously the unarmed master disarmed the officer & returned his sword. Again & again the officer tried to kill, using every subtle kata in his repertoire--but out of his empty mind the hermit each time invented a new way to disarm him.

The officer of course became his first disciple. Later, they learned how to dodge bullets. We might contemplate some form of metadrama meant to capture a taste of this performance, which gave rise to a wholly new art, a totally non-violent way of fighting--war without murder, "the sword of life" rather than death.

A conspiracy of artists, anonymous as any mad bombers, but aimed toward an act of gratuitous generosity rather than violence--at the millennium rather than the apocalypse--or rather, aimed at a present moment of aesthetic shock in the service of realization & liberation.

Art tells gorgeous lies that come true.

Is it possible to create a SECRET THEATER in which both artist & audience have completely disappeared--only to re-appear on another plane, where life & art have become the same thing, the pure giving of gifts?

(Note: The "Salon Apocalypse" was organized by Sharon Gannon in July, 1986.)

II. Murder--War--Famine--Greed

THE MANICHEES & CATHARS believed that the body can be spiritualized--or rather, that the body merely contaminates pure spirit & must be utterly rejected. The Gnostic perfecti (radical dualists) starved themselves to death to escape the body & return to the pleroma of pure light. So: to evade the evils of the flesh--murder, war, famine, greed--paradoxically only one path remains: murder of one's own body, war on the flesh, famine unto death, greed for salvation.

The radical monists however (Ismailis, Ranters, Antinomians) consider that body & spirit are one, that the same spirit which pervades a black stone also infuses the flesh with its light; that all lives & all is life.

"Things are what they are spontaneously...everything is natural...all in motion as if there were a True Lord to move them--but if we seek for evidence of this lord we fail to find any." (Kuo Hsiang)

Paradoxically, the monist path also cannot be followed without some sort of "murder, war, famine, greed": the transformation of death into life (food, negentropy)--war against the Empire of Lies--"fasting of the soul," or renunciation of the Lie, of all that is not life--& greed for life itself, the absolute power of desire.

Even more: without knowledge of the darkness ("carnal knowledge") there can exist no knowledge of the light ("gnosis"). The two knowledges are not merely complementary: say rather identical, like the same note played in different octaves.

Heraclitus claims that reality persists in a state of "war." Only clashing notes can make harmony. ("Chaos is the sum of all orders.") Give each of these four terms a different mask of language (to call the Furies "The Kindly Ones" is not mere euphemism but a way of uncovering yet more meaning). Masked, ritualized, realized as art, the terms take on their dark beauty, their "Black Light."

Instead of murder say the hunt, the pure paleolithic economy of all archaic and non-authoritarian tribal society--"venery," both the killing & eating of flesh & the way of Venus, of desire. Instead of war say insurrection, not the revolution of classes & powers but of the eternal rebel, the dark one who uncovers light. Instead of greed say yearning, unconquerable desire, mad love. And then instead of famine, which is a kind of mutilation, speak of wholeness, plenty, superabundance, generosity of the self which spirals outward toward the Other.

Without this dance of masks, nothing will be created. The oldest mythology makes Eros the firstborn of Chaos. Eros, the wild one who tames, is the door through which the artist returns to Chaos, the One, and then re-returns, comes back again, bearing one of the patterns of beauty. The artist, the hunter, the warrior: one who is both passionate and balanced, both greedy & altruistic to the utmost extreme. We must be saved from all salvations which save us from ourselves, from our animal which is also our anima, our very lifeforce, as well as our animus, our animating self-empowerment, which may even manifest as anger & greed. BABYLON has told us that our flesh is filth--with this device & the promise of salvation it enslaved us. But--if the flesh is already "saved," already light--if even consciousness itself is a kind of flesh, a palpable & simultaneous living aether-- then we need no power to intercede for us. The wilderness, as Omar says, is paradise even now.

The true proprietorship of murder lies with the Empire, for only freedom is complete life. War is Babylonian as well--no free person will die for another's aggrandizement. Famine comes into existence only with the civilization of the saviors, the priest-kings--wasn't it Joseph who taught Pharaoh to speculate in grain futures? Greed--for land, for symbolic wealth, for power to deform others' souls & bodies for their own salvation--greed too arises not from "Nature nature-ing," but from the damming up & canalization of all energies for the Empire's Glory. Against all this, the artist possesses the dance of masks, the total radicalization of language, the invention of a "Poetic Terrorism" which will strike not at living beings but at malign ideas, dead-weights on the coffin-lid of our desires. The architecture of suffocation and paralysis will be blown up. only by our total celebration of everything-- even darkness.

--Summer Solstice, 1986

COMMUNIQUE #7

Psychic Paleolithism & High Technology: A Position Paper

JUST BECAUSE THE A.O.A. talks about "Paleolithism" all the time, don't get the idea we intend to bomb ourselves back to the Stone Age.

We have no interest in going "back to the land" if the deal includes the boring life of a shit-kicking peasant--nor do we want "tribalism" if it comes with taboos, fetishes & malnutrition. We have no quarrel with the concept of culture--including technology; for us the problem begins with civilization.

What we like about Paleolithic life has been summed up by the Peoples-Without-Authority School of anthropology: the elegant laziness of hunter/gatherer society, the 2-hour workday, the obsession with art, dance, poetry & amorousness, the "demo-cratization of shamanism," the cultivation of perception--in short, culture.

What we dislike about civilization can be deduced from the following progression: the "Agricultural Revolution"; the emergence of caste; the City & its cult of hieratic control ("Babylon"); slavery; dogma; imperialism ("Rome"). The suppression of sexuality in "work" under the aegis of "authori-ty." "The Empire never ended."

A psychic paleolithism based on High-Tech--post-agricultural, post-industrial, "Zerowork," nomadic (or "Rootless Cosmopoli-tan")--a Quantum Paradigm Society--this constitutes the ideal vision of the future according to Chaos Theory as well as "Futurology" (in the Robert Anton Wilson-T. Leary sense of the term).

As for the present: we reject all collaboration with the Civiliza-tion of Anorexia & Bulimia, with people so ashamed of never suffering that they invent hair shirts for themselves & others--or those who gorge without compassion & then spew the vomit of their suppressed guilt in great masochistic bouts of jogging & dieting. All our pleasures & self-disciplines belong to us by Nature--we never deny ourselves, we never give up anything; but some things have given up on us & left us, because we are too large for them. I am both caveman & starfaring mutant, con-man & free prince. Once an Indian Chief was invited to the White House for a banquet. As the food passed round, the Chief heaped his plate to the max, not once but three times. At last the honky sitting next to him says, "Chief, heh-heh, don't you think that's a little too much?" "Ugh," the Chief replies, "little too much just right for Chief!"

Nevertheless, certain doctrines of "Futurology" remain problematic. For example, even if we accept the liberatory potential of such new technologies as TV, computers, robotics, Space exploration, etc., we still see a gap between potentiality & actualization. The banalization of TV, the yuppification of computers & the militarization of Space suggest that these technologies in themselves provide no "determined" guarantee of their liberatory use.

Even if we reject the Nuclear Holocaust as just another Spectacular Diversion orchestrated to distract our attention from real problems, we must still admit that "Mutual Assured Destruction" & "Pure War" tend to dampen our enthusiasm for certain aspects of the High-Tech Adventure. Ontological Anarchy retains its affection for Luddism as a tactic: if a given technology, no matter how admirable in potentia (in the future), is used to oppress me here & now, then I must either wield the weapon of sabotage or else seize the means of production (or perhaps more importantly the means of communication). There is no humanity without techne--but there is no techne worth more than my humanity.

We spurn knee-jerk anti-Tech anarchism--for ourselves, at least (there exist some who enjoy farming, or so one hears)--and we reject the concept of the Technological Fix as well. For us all forms of determinism appear equally vapid--we're slaves of neither our genes nor our machines. What is "natural" is what we imagine & create. "Nature has no Laws--only habits."

Life for us belongs neither to the Past--that land of famous ghosts hoarding their tarnished grave-goods--nor to the Future, whose bulbbrained mutant citizens guard so jealously the secrets of immortality, faster-than-light flight, designer genes & the withering of the State. Aut nunc aut nihil. Each moment contains an eternity to be penetrated--yet we lose ourselves in visions seen through corpses' eyes, or in nostalgia for unborn perfections.

The attainments of my ancestors & descendants are nothing more to me than an instructive or amusing tale--I will never call them my betters, even to excuse my own smallness. I print for myself a license to steal from them whatever I need--psychic paleolithism or high-tech--or for that matter the gorgeous detritus of civilization itself, secrets of the Hidden Masters, pleasures of frivolous nobility & la vie boheme.

La decadence, Nietzsche to the contrary notwithstanding, plays as deep a role in Ontological Anarchy as health--we take what we want of each. Decadent aesthetes do not wage stupid wars nor submerge their consciousness in microcephalic greed & resentment. They seek adventure in artistic innovation & non-ordinary sexuality rather than in the misery of others. The A.O.A. admires & emulates their sloth, their disdain for the stupidity of normalcy, their expropriation of aristocratic sensibilities. For us these qualities harmonize paradoxically with those of the Old Stone Age & its overflowing health, ignorance of hierarchy, cultivation of virtu rather than Law. We demand decadence without sickness, & health without boredom!

Thus the A.O.A. gives unqualified support to all indigenous & tribal peoples in their struggle for complete autonomy--& at the same time, to the wildest, most Spaced-out speculations & demands of the Futurologists. The paleolithism of the future (which for us, as mutants, already exists) will be achieved on a grand scale only through a massive technology of the Imagina-tion, and a scientific paradigm which reaches beyond Quantum Mechanics into the realm of Chaos Theory & the hallucinations of Speculative Fiction.

As Rootless Cosmopolitans we lay claim to all the beauties of the past, of the orient, of tribal societies--all this must & can be ours, even the treasuries of the Empire: ours to share. And at the same time we demand a technology which transcends agriculture, industry, even the simultaneity of electricity, a hardware that intersects with the wetware of consciousness, that embraces the power of quarks, of particles travelling backward in time, of quasars & parallel universes.

The squabbling ideologues of anarchism & libertarianism each prescribe some utopia congenial to their various brands of tunnel-vision, ranging from the peasant commune to the L-5 Space City. We say, let a thousand flowers bloom--with no gardener to lop off weeds & sports according to some moralizing or eugenical scheme. The only true conflict is that between the authority of the tyrant & the authority of the realized self-- all else is illusion, psychological projection, wasted verbiage.

In one sense the sons & daughters of Gaia have never left the paleolithic; in another sense, all the perfections of the future are already ours. Only insurrection will "solve" this paradox-- only the uprising against false consciousness in both ourselves & others will sweep away the technology of oppression & the poverty of the Spectacle. In this battle a painted mask or shaman's rattle may prove as vital as the seizing of a communications satellite or secret computer network.

Our sole criterion for judging a weapon or a tool is its beauty. The means already are the end, in a certain sense; the insurrection already is our adventure; Becoming IS Being. Past & future exist within us & for us, alpha & omega. There are no other gods before or after us. We are free in TIME--and will be free in SPACE as well.

(Thanx to Hagbard Celine the Sage of Howth & Environs)

COMMUNIQUE #8

Chaos Theory & the Nuclear Family

SUNDAY IN RIVERSIDE PARK the Fathers fix their sons in place, nailing them magically to the grass with baleful ensorcelling stares of milky camaraderie, & force them to throw baseballs back & forth for hours. The boys almost appear to be small St Sebastians pierced by arrows of boredom.

The smug rituals of family fun turn each humid Summer meadow into a Theme Park, each son an unwitting allegory of Father's wealth, a pale representation 2 or 3 times removed from reality: the Child as metaphor of Something-or-other.

And here I come as dusk gathers, stoned on mushroom dust, half convinced that these hundreds of fireflies arise from my own consciousness--Where have they been all these years? why so many so suddenly?--each rising in the moment of its incandescence, describing quick arcs like abstract graphs of the energy in sperm.

"Families! misers of love! How I hate them!" Baseballs fly aimlessly in vesper light, catches are missed, voices rise in peevish exhaustion. The children feel sunset encrusting the last few hours of doled-out freedom, but still the Fathers insist on stretching the tepid postlude of their patriarchal sacrifice till dinnertime, till shadows eat the grass.

Among these sons of the gentry one locks gazes with me for a moment--I transmit telepathically the image of sweet license, the smell of TIME unlocked from all grids of school, music lessons, summer camps, family evenings round the tube, Sundays in the Park with Dad--authentic time, chaotic time.

Now the family is leaving the Park, a little platoon of dissatisfaction. But that one turns & smiles back at me in complicity-- "Message Received"--& dances away after a firefly, buoyed up by my desire. The Father barks a mantra which dissipates my power.

The moment passes. The boy is swallowed up in the pattern of the week--vanishes like a bare-legged pirate or Indian taken prisoner by missionaries. The Park knows who I am, it stirs under me like a giant jaguar about to wake for nocturnal meditation. Sadness still holds it back, but it remains untamed in its deepest essence: an exquisite disorder at the heart of the city's night.

COMMUNIQUE #9

Double-Dip Denunciations

I. Xtianity

AGAIN & AGAIN WE hope that attitudinizing corpse has finally breathed its last rancorous sigh & floated off to its final pumpkinification. Again & again we imagine the defeat of that obscene flayed death-trip bogey nailed to the walls of all our waiting rooms, never again to whine at us for our sins...

but again & again it resurrects itself & comes creeping back to haunt us like the villain of some nth rate snuff-porn splatter film--the thousandth re-make of Night of the Living Dead-- trailing its snail-track of whimpering humiliation...just when you thought it was safe in the unconscious...it's JAWS for JESUS. Look out! Hardcore Chainsaw Baptists!

and the Leftists, nostalgic for the Omega Point of their dialectical paradise, welcome each galvanized revival of the putrescent creed with coos of delight: Let's dance the tango with all those marxist bishops from Latin America--croon a ballad for the pious Polish dockworkers--hum spirituals for the latest afro-Methodist presidential hopeful from the Bible Belt...

The A.O.A. denounces Liberation Theology as a conspiracy of stalinist nuns--the Whore of Babylon's secret scarlet deal with red fascism in the tropics. Solidarnosc? The Pope's Own Labor Union--backed by the AFL/CIO, the Vatican Bank, the Freemason Lodge Propaganda Due, and the Mafia. And if we ever voted we'd never waste that empty gesture on some Xtian dog, no matter what its breed or color.

As for the real Xtians, those bored-again self-lobotomized bigots, those Mormon babykillers, those Star Warriors of the Slave Morality, televangelist blackshirts, zombie squads of the Blessed Virgin Mary (who hovers in a pink cloud over the Bronx spewing hatred, anathema, roses of vomit on the sexuality of children, pregnant teenagers & queers)...

As for the genuine death-cultists, ritual cannibals, Armageddon-freaks--the Xtian Right--we can only pray that the RAPTURE WILL COME & snatch them all up from behind the steering wheels of their cars, from their lukewarm game shows & chaste beds, take them all up into heaven & let us get on with human life.

II. Abortionists & Anti-abortionists

52

REDNECKS WHO BOMB ABORTION clinics belong in the same grotesque category of vicious stupidity as bishops who prattle Peace & yet condemn all human sexuality. Nature has no laws ("only habits"), & all law is unnatural. Everything belongs to the sphere of personal/imaginal morality--even murder.

However, according to Chaos Theory, it does not follow that we are obliged to like & approve of murder--or abortion. Chaos would enjoy seeing every bastard love-child carried to term & birthed; sperm & egg alone are mere lovely secretions, but combined as DNA they become potential consciousness, negentropy, joy.

If "meat is murder!" as the Vegans like to claim, what pray tell is abortion? Those totemists who danced to the animals they hunted, who meditated to become one with their living food & share its tragedy, demonstrated values far more humane than the average claque of "pro-Choice" feminoid liberals.

In every single "issue" cooked up for "debate" in the pattern-book of the Spectacle, both sides are invariably full of shit. The "abortion issue" is no exception.

COMMUNIQUE #10

Plenary Session Issues New Denunciations--Purges Expected

TO OFFSET ANY STICKY karma we might have acquired thru our pulpit-thumping sermonette against Xtians & other end-of-the-world creeps (see last ish) & just to set the record straight: the A.O.A. also denounces all born-again knee-jerk atheists & their frowsy late-Victorian luggage of scientistic vulgar material-ism. ///// We applaud all anti-Xtian sentiment, of course --& all attacks on all organized religions. But... to hear some anarch-ists talk you'd think the sixties never happened and no one ever dropped LSD. ///// As for the scientists themselves, the Alice-like madnesses of Quantum & Chaos Theory have driven the best of them towards taoism & vedanta (not to mention dada)--& yet if you read The Match or Freedom you might imagine science was embalmed with Prince Kropotkin--& "religion" with Bishop Ussher. ///// Of course one despises the Aquarian brownshirts, the kind of gurus lauded recently in the New York Times for their contributions to Big Business, the franchise-granting yuppie zombie cults, the anorexic metaphysics of New Age banality...but OUR esotericism remains undefiled by these mediocre money-changers & their braindead minions. ///// The heretics & antinomian mystics of Orient & Occident have developed systems based on inner liberation. Some of these systems are tainted with religious mysticism & even social reaction--others seem more purely radical or "psychological"--& some even crystallize into revolutionary movements (millena-rian Levellers, Assassins, Yellow Turban Taoists, etc.) Whatever their flaws they possess certain magical weapons which anarchism sorely lacks: (1) A sense of the meta-rational ("metanoia"), ways to go beyond laminated thinking into smooth (or nomadic or "chaotic") thinking & perception; (2) an actual definition of self-realized or liberated consciousness, a positive description of its structure, & techniques for approach-ing it; (3) a coherent archetypal view of epistemology--that is, a way of knowing (about history, for example) that utilizes hermeneutic phenomenology to uncover patterns of meaning (something like the Surrealists' "Paranoia Criticism"); (4) a teaching on sexuality (in the "tantrik" aspects of various Paths) that assigns value to pleasure rather than self-denial, not only for its own sake but as a vehicle of enhanced awareness or "liberation"; (5) an attitude of celebration, what might be called a "Jubilee concept," a cancelling of psychic debt thru some inherent generosity in reality itself; (6) a language (including gesture, ritual, intentionality) with which to animate &

communicate these five aspects of cognition; and (7) a silence. ///// It's no surprise to discover how many anarchists are ex-Catholics, defrocked priests or nuns, former altar boys, lapsed born-again baptists or even ex-Shiite fanatics. Anarchism offers up a black (& red) Mass to de-ritualize all spook-haunted brains--a secular exorcism--but then betrays itself by cobbling together a High Church of its own, all cobwebby with Ethical Humanism, Free Thought, Muscular Atheism, & crude Fundamentalist Cartesian Logic. ///// Two decades ago we began the project of becoming Rootless Cosmopolitans, determined to sift the detritus of all tribes, cultures & civilizations (including our own) for viable fragments--& to synthesize from this mess of potsherds a living system of our own--lest (as Blake warned) we become slaves to someone else's. ///// If some Javanese sorcerer or Native American shaman possesses some precious fragment I need for my own "medicine pouch," should I sneer & quote Bakunin's line about stringing up priests with bankers' guts? or should I remember that anarchy knows no dogma, that Chaos cannot be mapped--& help myself to anything not nailed down? ///// The earliest definitions of anarchy are found in the Chuang Tzu & other taoist texts; "mystical anarchism" boasts a hoarier pedigree than the Greco-Rationalist variety. When Nietzsche spoke of the "Hyperboreans" I think he foretold us, who have gone beyond the death of God--& the rebirth of the Goddess--to a realm where spirit & matter are one. Every manifestation of that hierogamy, every material thing & every life, becomes not only "sacred" in itself but also symbolic of its own "divine essence." ///// Atheism is nothing but the opiate of The Masses (or rather, their self-chosen champions)--& not a very colorful or sexy drug. If we are to follow Baudelaire's advice & "be always intoxicated," the A.O.A. would prefer something more like mushrooms, thank you. Chaos is the oldest of the gods--& Chaos never died.

COMMUNIQUE #11

Special Holiday Season Food Issue Rant: Turn Off the Lite!

THE ASSOCIATION FOR ONTOLOGICAL ANARCHY calls for a boycott of all products marketed under the Shibboleth of LITE-- beer, meat, local candy, cosmetics, music, pre-packaged "lifestyles," whatever.

The concept of LITE (in Situ-jargon) unfolds a complex of symbolism by which the Spectacle hopes to recuperate all revulsion against its commodification of desire. "Natural," "organic," "healthy" produce is designed for a market sector of mildly dissatisfied consumers with mild cases of future- shock & mild yearnings for a tepid authenticity. A niche has been prepared for you, softly illumined with the illusions of simplicity, cleanliness, thinness, a dash of asceticism & self-denial. Of course, it costs a little more...after all, LITEness was not designed for poor hungry primitivos who still think of food as nourishment rather than decor. It has to cost more--otherwise you wouldn't buy it.

The American Middle Class (don't quibble; you know what I mean) falls naturally into opposite but complementary factions: the Armies of Anorexia & Bulimia. Clinical cases of these diseases represent only the psychosomatic froth on a wave of cultural pathology, deep, diffused & largely unconscious. The Bulimics are those yupped-out gentry who gorge on margharitas & VCRs, then purge on LITE food, jogging, or (an)aerobic jiggling. The Anorexics are the "lifestyle" rebels, ultra-food-faddists, eaters of algae, joyless, dispirited & wan--but smug in their puritanical zeal & their designer hair-shirts. Grotesque junk food simply represents the flip-side of ghoulish "health food":-- nothing tastes like anything but woodchips or additives--it's all either boring or carcinogenic--or both--& it's all incredibly stupid.

Food, cooked or raw, cannot escape from symbolism. It is, & also simultaneously represents that which it is. All food is soul food; to treat it otherwise is to court indigestion, both chronic & metaphysical.

But in the airless vault of our civilization, where nearly every experience is mediated, where reality is strained through the deadening mesh of consensus-perception, we lose touch with food as nourishment; we begin to construct for ourselves personae based on what we consume, treating products as projections of our yearning for the authentic.

The A.O.A. sometimes envisions CHAOS as a cornucopia of continual creation, as a sort of geyser of cosmic generosity; therefore we refrain from advocating any specific diet, lest we offend against the Sacred Multiplicity & the Divine Subjectivity. We're not about to hawk you yet another New Age prescription for perfect health (only the dead are perfectly healthy); we interest ourselves in life, not "lifestyles."

True lightness we adore, & rich heaviness delights us in its season. Excess suits us to perfection, moderation pleases us, & we have learned that hunger can be the finest of all spices. Everything is light, & the lushest flowers grow round the privy. We dream of phalanstery tables & bolo'bolo cafes where every festive collective of diners will share the individual genius of a Brillat-Savarin (that saint of taste).

Shaykh Abu Sa'id never saved money or even kept it overnight-- therefore, whenever some patron donated a heavy purse to his hospice, the dervishes celebrated with a gourmet feast; & on other days, all went hungry. The point was to enjoy both states, full & empty...

LITE parodies spiritual emptiness & illumination, just as McDonald's travesties the imagery of fullness & celebration. The human spirit (not to mention hunger) can overcome & transcend all this fetishism--joy can erupt even at Burger King, & even LITE beer may hide a dose of Dionysus. But why should we have to struggle against this garbagy tide of cheap rip-off ticky-tack, when we could be drinking the wine of paradise even now under our own vine & fig tree?

Food belongs to the realm of everyday life, the primary arena for all insurrectionary self-empowerment, all spiritual self-enhancement, all seizing-back of pleasure, all revolt against the Planetary Work Machine & its imitation desires. Far be it from us to dogmatize; the Native American hunter might fuel his happiness with fried squirrel, the anarcho-taoist with a handful of dried apricots. Milarepa the Tibetan, after ten years of nettle-soup, ate a butter cake & achieved enlightenment. The dullard sees no eros in fine champagne; the sorcerer can fall intoxicated on a glass of water.

Our culture, choking on its own pollutants, cries out (like the dying Goethe) for "More LITE!"--as if these polyunsaturated effluents could somehow assuage our misery, as if their bland weightless tasteless characterlessness could protect us from the gathering dark.

No! This last illusion finally strikes us as too cruel. We are forced against our own slothful inclinations to take a stand & protest. Boycott! Boycott! TURN OFF THE LITE!

Appendix: Menu For An Anarchist Black Banquet (veg & non-veg)

Caviar & blinis; Hundred year old eggs; Squid & rice cooked in ink; Eggplants cooked in their skins with black pickled garlic; Wild rice with black walnuts & black mushrooms; Truffles in black butter; Venison marinated in port, charcoal grilled, served on pumpernickel slices & garnished with roast chestnuts. Black Russians; Guiness-&-champagne; Chinese black tea. Dark chocolate mousse, Turkish coffee, black grapes, plums, cherries, etc.

SPECIAL HALLOWEEN COMMUNIQUE

Black Magic as Revolutionary Action

PREPARE AN INK OF pure & genuine saffron mixed with rose-water, adding if possible some blood from a black rooster. In a quiet room furnish an altar with a bowl of the ink, a pen with an iron nib, 7 black candles, an incense burner, & some benzoin. The charm may be written on virgin paper or parchment. Draw the diagram at 4 p.m. on a Wednesday, facing North. Copy the 7-headed diagram (see illustration) without lifting the pen from the paper, in one smooth operation, holding your breath & pressing your tongue to the roof of your mouth. This is the Barisan Laksamana, or King of the Djinn. Then draw the Solomon's Seal (a star representing a 5-headed djinn) & other parts of the diagram. Above Solomon's Seal write the name of the individual or institution to be cursed. Now hold the paper in the benzoin fumes, & invoke the white & black djinn within yourself:

Bismillah ar-Rahman ar-Rahim
as-salaam alikum
O White Djinn, Radiance of Mohammad

king of all spirits within me
O Black Djinn, shadow of myself
AWAY, destroy my enemy
--and if you do not
then be considered a traitor to Allah
--by virtue of the charm
La illaha ill'Allah

Mohammad ar-Rasul Allah

If the curse is to be aimed at an individual oppressor, a wax doll may be prepared & the charm inserted (see illustration).

Seven needles are then driven downward into the top of the head, thru the left & right armpits, left & right hips, & thru the lips or nostrils. Wrap the doll in a white shroud & bury it in the ground where the enemy is sure to walk over it, meanwhile enlisting the aid of local earth spirits:

Bismillah ar-Rahman ar-Rahim
O Earth Djinn, Dirt-spirit
O Black Djinn living underground
listen, vampire of the soil
I order you to mark & destroy
the body & soul of
Heed my orders
for I am the true & original sorcerer
by virtue of the charm
la illaha ill'Allah

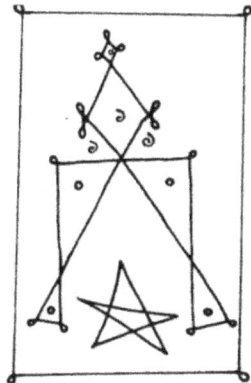

Mohammad ar-Rasul Allah

If however the curse is intended for an institution or company, assemble the following items: a hard-boiled egg, an iron nail, &

3 iron pins (stick nail & needles into egg); dried scorpion, lizard &/or beetles; a small chamois bag containing graveyard dirt, magnetized iron fillings, asafoetida & sulphur, & tied with a red ribbon. Sew the charm into yellow silk & seal it with red wax. Place all these things in a wide-necked bottle, cork it, & seal it with wax.

The bottle may now be carefully packaged & sent by mail to the target institution--for example a Xtian televangelist show, the New York Post, the MUZAK company, a school or college--along with a copy of the following statement (extra copies may be mailed to individual employees, &/or posted surreptitiously around the premises):

Malay Black Djinn Curse

These premises have been cursed by black sorcery. The curse has been activated according to correct rituals. This institution is cursed because it has oppressed the Imagination & defiled the Intellect, degraded the arts toward stupefaction, spiritual slavery, propaganda for State & Capital, puritanical reaction, unjust profits, lies & aesthetic blight. The employees of this institution are now in danger. No ind ividual has been cursed, but the place itself has been infec ted with ill fortune & malignancy. Those who do not wake up & quit, or begin sabotaging the workplace, will gradually fa ll under the effect of this sorcery. Removing or destroying the implement of sorcery will do no good. It has been seen i n this place, & this place is cursed. Reclaim your humanity & revolt in the name of the Imagination--or else be judged (in the mirror of this charm) an enemy of the human race.

We suggest "taking credit" for this action in the name of some other offensive cultural institution, such as the American Poetry Society or the Women's Anti-Porn Crusade (give full address).

We also suggest, in order to counter-balance the effect on yourself of calling up the personal black djinn, that you send a magical blessing to someone or some group you love &/or admire. Do this anonymously, & make the gift beautiful. No precise ritual need be followed, but the imagery should be allowed to spring from the well of consciousness in an intuitive/spontaneous meditational state. Use sweet incense, red & white candles, hard candy, wine, flowers, etc. If possible include real silver, gold, or jewels in the gift.

This how-to-do-it manual on the Malay Black Djinn Curse has been prepared according to authentic & complete ritual by the Cultural Terrorism Committee of the inner Adept Chamber of the HMOCA ("Third Paradise"). We are Nizari-Ismaili Esotericists; that is, Shiite heretics & fanatics who trace our spiritual line to Hassan-i Sabbah through Aladdin Mohammad III "the Madman," seventh & last Pir of Alamut (& not through the line of the Aga Khans). We espouse radical monism & pure antinomianism, & oppose all forms of law & authority, in the name of Chaos.

At present, for tactical reasons, we do not advocate violence or sorcery against individuals. We call for actions against institutions & ideas--art-sabotage & clandestine propaganda (including ceremonial magic & "tantrik pornography")--and especially against the poisonous media of the Empire of Lies. The Black Djinn Curse represents only a first step in the campaign of Poetic Terrorism which--we trust--will lead to other less subtle forms of insurrection.

SPECIAL COMMUNIQUE

A.O.A. Announces Purges in Chaos Movement

CHAOS THEORY MUST OF course flow impurely. "Lazy yokel plows a crooked furrow." Any attempt to precipitate a crystal of ideology would result in flawed rigidities, fossilizations, armorings & drynesses which we would like to renounce, along with all "purity." Yes, Chaos revels in a certain abandoned formlessness not unlike the erotic messiness of those we love for their shattering of habit & their unveiling of mutability. Nevertheless this looseness does not imply that Chaos Theory must accept every leech that attempts to attach itself to our sacred membranes. Certain definitions or deformations of Chaos deserve denunciation, & our dedication to divine disorder need not deter us from trashing the traitors & rip-off artists & psychic vampires now buzzing around Chaos under the impression that it's trendy. We propose not an Inquisition in the name of our definitions, but rather a duel, a brawl, an act of violence or emotional repugnance, an exorcism. First we'd like to define & even name our enemies. (1) All those death-heads & mutilation artists who associate Chaos exclusively with misery, negativity & a joyless pseudo-libertinism--those who think "beyond good & evil" means doing evil--the S/M intellectuals, crooners of the apocalypse--the new Gnostic Dualists, world-haters & ugly nihilists. (2) All those scientists selling Chaos either as a force for destruction (e.g. particle-beam weapons) or as a mechanism for enforcing order, as in the use of Chaos math in statistical sociology and mob control. An attempt will be made to discover names and addresses in this category. (3) All those who appropriate Chaos in the cause of some New Age scam. Of course we have no objection to your giving us all your money, but we'll tell you up front: we'll use it to buy dope or fly to Morocco. You can't sell water by the river; Chaos is that materia of which the alchemists spoke, which fools value more highly than gold even tho it may be found on any dungheap. The chief enemy in this category is Werner Erhardt, founder of est, who is now bottling "Chaos" & trying to franchise it to the Yuppoids.

Second, we will list some of our friends, in order to give an idea of the disparate trends in Chaos Theory we enjoy: Chaotica, the imaginal autonomous zone discovered by Feral Faun (a.k.a. Feral Ranter); the Academy of Chaotic Arts of Tundra Wind; Joel Birnoco's magazine KAOS; Chaos Inc., a newsletter connected to the work of Ralph Abraham, a leading Chaos scientist; the Church of Eris; Discordian Zen; the Moorish Orthodox Church; certain clenches of the Church of the SubGenius; the Sacred Jihad of Our Lady of Perpetual Chaos; the writers associated with "type-3 anarchism" & journals like Popular Reality; etc. The battle lines are drawn. Chaos is not entropy, Chaos is not death, Chaos is not a commodity. Chaos is continual creation. Chaos never died.

POST-ANARCHISM ANARCHY

THE ASSOCIATION FOR ONTOLOGICAL ANARCHY gathers in conclave, black turbans & shimmering robes, sprawled on shirazi carpets sipping bitter coffee, smoking long chibouk & sibsi. QUESTION: What's our position on all these recent defections & desertions from anarchism (esp. in California-Land): condemn or condone? Purge them or hail them as advance-guard? Gnostic elite...or traitors?

Actually, we have a lot of sympathy for the deserters & their various critiques of anarchISM. Like Sinbad & the Horrible Old Man, anarchism staggers around with the corpse of a Martyr magically stuck to its shoulders--haunted by the legacy of failure & revolutionary masochism--stagnant backwater of lost history.

Between tragic Past & impossible Future, anarchism seems to lack a Present--as if afraid to ask itself, here & now, WHAT ARE MY TRUE DESIRES?--& what can I DO before it's too late?...Yes, imagine yourself confronted by a sorcerer who stares you down balefully & demands, "What is your True Desire?" Do you hem & haw, stammer, take refuge in ideological platitudes? Do you possess both Imagination & Will, can you both dream & dare--or are you the dupe of an impotent fantasy?

Look in the mirror & try it...(for one of your masks is the face of a sorcerer)...

The anarchist "movement" today contains virtually no Blacks, Hispanics, Native Americans or children...even tho in theory such genuinely oppressed groups stand to gain the most from any anti-authoritarian revolt. Might it be that anarchISM offers no concrete program whereby the truly deprived might fulfill (or at least struggle realistically to fulfill) real needs & desires?

If so, then this failure would explain not only anarchism's lack of appeal to the poor & marginal, but also the disaffection & desertions from within its own ranks. Demos, picket-lines & reprints of 19th century classics don't add up to a vital, daring conspiracy of self-liberation. If the movement is to grow rather than shrink, a lot of deadwood will have to be jettisoned & some risky ideas embraced.

The potential exists. Any day now, vast numbers of americans are going to realize they're being force-fed a load of reactionary boring hysterical artificially-flavored crap. Vast chorus of groans, puking & retching...angry mobs roam the malls, smashing & looting...etc., etc. The Black Banner could provide a focus for the outrage & channel it into an insurrection of the Imagination. We could pick up the struggle where it was dropped by Situationism in '68 & Autonomia in the seventies, & carry it to the next stage. We could have revolt in our times--& in the process, we could realize many of our True Desires, even if only for a season, a brief Pirate Utopia, a warped free-zone in the old Space/Time continuum.

If the A.O.A. retains its affiliation with the "movement," we do so not merely out of a romantic predilection for lost causes--or not entirely. Of all "political systems," anarchism (despite its flaws, & precisely because it is neither political nor a system) comes closest to our understanding of reality, ontology, the nature of being. As for the deserters...we agree with their critiques, but note that they seem to offer no new powerful alternatives. So for the time being we prefer to concentrate on changing anarchism from within. Here's our program, com-rades:

Work on the realization that psychic racism has replaced overt discrimination as one of the most disgusting aspects of our society. Imaginative participation in other cultures, esp. those we live with.

Abandon all ideological purity. Embrace "Type-3" anarchism (to use Bob Black's pro-tem slogan): neither collectivist nor individualist. Cleanse the temple of vain idols, get rid of the Horrible Old Men, the relics & martyrologies.

Anti-work or "Zerowork" movement extremely important, including a radical & perhaps violent attack on Education & the serfdom of children.

Develop american samizdat network, replace outdated publishing/propaganda tactics. Pornography & popular entertainment as vehicles for radical re-education.

In music the hegemony of the 2/4 & 4/4 beat must be overth-rown. We need a new music, totally insane but life-affirming, rhythmically subtle yet powerful, & we need it NOW.

Anarchism must wean itself away from evangelical materialism & banal 2-dimensional 19th century scientism. "Higher states of consciousness" are not mere SPOOKS invented by evil priests. The orient, the occult, the tribal cultures possess techniques which can be "appropriated" in true anarchist fashion. Without "higher states of consciousness," anarchism ends & dries itself up into a form of misery, a whining complaint. We need a practical kind of "mystical anarchism," devoid of all New Age shit-&-shinola, & inexorably heretical & anti-clerical; avid for all new technologies of consciousness & metanoia--a democratization of shamanism, intoxicated & serene.

Sexuality is under assault, obviously from the Right, more subtly from the avant-pseud "post-sexuality" movement, & even more subtly by Spectacular Recuperation in media & advertising. Time for a major step forward in SexPol awareness, an explosive reaffirmation of the polymorphic eros--(even & especially in the face of plague & gloom)--a literal glorification of the senses, a doctrine of delight. Abandon all world-hatred & shame.

Experiment with new tactics to replace the outdated baggage of Leftism. Emphasize practical, material & personal benefits of radical networking. The times do not appear propitious for violence or militancy, but surely a bit of sabotage & imaginative disruption is never out of place. Plot & conspire, don't bitch & moan. The Art World in particular deserves a dose of "Poetic Terrorism."

The despatialization of post-Industrial society provides some benefits (e.g. computer networking) but can also manifest as a form of oppression (homelessness, gentrification, architectural depersonalization, the erasure of Nature, etc.) The communes of the sixties tried to circumvent these forces but failed. The question of land refuses to go away. How can we separate the concept of space from the mechanisms of control? The territorial gangsters, the Nation/States, have hogged the entire map. Who can invent for us a cartography of autonomy, who can draw a map that includes our desires?

AnarchISM ultimately implies anarchy--& anarchy is chaos. Chaos is the principle of continual creation...& Chaos never died.

--A.O.A. Plenary Session

March '87, NYC

BLACK CROWN & BLACK ROSE

Anarcho-Monarchism & Anarcho-Mysticism

IN SLEEP WE DREAM of only two forms of government--anarchy & monarchy. Primordial root consciousness understands no politics & never plays fair. A democratic dream? a socialist dream? Impossible.

Whether my REMs bring verdical near-prophetic visions or mere Viennese wish-fulfillment, only kings & wild people populate my night. Monads & nomads.

Pallid day (when nothing shines by its own light) slinks & insinuates & suggests that we compromise with a sad & lackluster reality. But in dream we are never ruled except by love or sorcery, which are the skills of chaotes & sultans.

Among a people who cannot create or play, but can only work, artists also know no choice but anarchy & monarchy. Like the dreamer, they must possess & do possess their own percep-tions, & for this they must sacrifice the merely social to a "tyrannical Muse." Art dies when treated "fairly." It must enjoy a caveman's wildness or else have its mouth filled with gold by some prince. Bureaucrats & sales personnel poison it, professors chew it up, & philosophers spit it out. Art is a kind of byzantine barbarity fit only for nobles & heathens. If you had known the sweetness of life as a poet in the reign of some venal, corrupt, decadent, ineffective & ridiculous Pasha or Emir, some Qajar shah, some King Farouk, some Queen of Persia, you would know that this is what every anarchist must want. How they loved poems & paintings, those dead luxurious fools, how they absorbed all roses & cool breezes, tulips & lutes! Hate their cruelty & caprice, yes--but at least they were human. The bureaucrats, however, who smear the walls of the mind with odorless filth--so kind, so gemutlich--who pollute the inner air with numbness--they're not even worthy of hate. They scarcely exist outside the bloodless Ideas they serve.

And besides: the dreamer, the artist, the anarchist--do they not share some tinge of cruel caprice with the most outrageous of moghuls? Can genuine life occur without some folly, some excess, some bouts of Heraclitan "strife"? We do not rule--but we cannot & will not be ruled.

In Russia the Narodnik-Anarchists would sometimes forge a ukase or manifesto in the name of the Czar; in it the Autocrat would complain that greedy lords & unfeeling officials had sealed him in his palace & cut him off from his beloved people. He would proclaim the end of serfdom & call on peasants & workers to rise in His Name against the government.

Several times this ploy actually succeeded in sparking revolts. Why? Because the single absolute ruler acts metaphorically as a mirror for the unique and utter absoluteness of the self. Each peasant looked into this glassy legend & beheld his or her own freedom--an illusion, but one that borrowed its magic from the logic of the dream.

A similar myth must have inspired the 17th century Ranters & Antinomians & Fifth Monarchy Men who flocked to the Jacobite standard with its erudite cabals & bloodproud conspiracies. The radical mystics were betrayed first by Cromwell & then by the Restoration--why not, finally, join with flippant cavaliers & foppish counts, with Rosicrucians & Scottish Rite Masons, to place an occult messiah on Albion's throne?

Among a people who cannot conceive human society without a monarch, the desires of radicals may be expressed in monarchical terms. Among a people who cannot conceive human existence without a religion, radical desires may speak the language of heresy.

Taoism rejected the whole of Confucian bureaucracy but retained the image of the Emperor-Sage, who would sit silent on his throne facing a propitious direction, doing absolutely nothing. In Islam the Ismailis took the idea of the Imam of the Prophet's Household & metamorphosed it into the Imam-of-one's-own-being, the perfected self who is beyond all Law & rule, who is atoned with the One. And this doctrine led them into revolt against Islam, to terror & assassination in the name of pure esoteric self-liberation & total realization.

Classical 19th century anarchism defined itself in the struggle against crown & church, & therefore on the waking level it considered itself egalitarian & atheist. This rhetoric however obscures what really happens: the "king" becomes the "anarchist," the "priest" a "heretic." In this strange duet of mutability the politician, the democrat, the socialist, the rational ideologue can find no place; they are deaf to the music & lack all sense of rhythm. Terrorist & monarch are archetypes; these others are mere functionaries.

Once anarch & king clutched each other's throats & waltzed a totentanz--a splendid battle. Now, however, both are relegated to history's trashbin--has-beens, curiosities of a leisurely & more cultivated past. They whirl around so fast that they seem to meld together...can they somehow have become one thing, a Siamese twin, a Janus, a freakish unity? "The sleep of Reason..." ah! most desirable & desirous monsters!

Ontological Anarchy proclaims flatly, bluntly, & almost brainlessly: yes, the two are now one. As a single entity the anarch/king now is reborn; each of us the ruler of our own flesh, our own creations--and as much of everything else as we can grab & hold.

Our actions are justified by fiat & our relations are shaped by treaties with other autarchs. We make the law for our own domains--& the chains of the law have been broken. At present perhaps we survive as mere Pretenders--but even so we may seize a few instants, a few square feet of reality over which to impose our absolute will, our royaume. L'etat, c'est moi.

If we are bound by any ethic or morality it must be one which we ourselves have imagined, fabulously more exalted & more liberating than the "moralic acid" of puritans & humanists. "Ye are as gods"--"Thou art That."

The words monarchism & mysticism are used here in part simply pour epater those egalito-atheist anarchists who react with pious horror to any mention of pomp or superstition-mongering. No champagne revolutions for them!

Our brand of anti-authoritarianism, however, thrives on baroque paradox; it favors states of consciousness, emotion & aesthetics over all petrified ideologies & dogma; it embraces multitudes & relishes contradictions. Ontological Anarchy is a hobgoblin for BIG minds. The translation of the title (& key term) of Max Stirner's magnum opus as The Ego & Its Own has led to a subtle misinterpretation of "individualism." The English-Latin word ego comes freighted & weighed with freudian & protestant baggage. A careful reading of Stirner suggests that The Unique & His Own-ness would better reflect his intentions, given that he never defines the ego in opposition to libido or id, or in opposition to "soul" or "spirit." The Unique (der Einzige) might best be construed simply as the individual self.

Stirner commits no metaphysics, yet bestows on the Unique a certain absoluteness. In what way then does this Einzige differ from the Self of Advaita Vedanta? Tat tvam asi: Thou (individual Self) art That (absolute Self).

Many believe that mysticism "dissolves the ego." Rubbish. Only death does that (or such at least is our Sadducean assumption). Nor does mysticism destroy the "carnal" or "animal" self--which would also amount to suicide. What mysticism really tries to surmount is false consciousness, illusion, Consensus Reality, & all the failures of self that accompany these ills. True mysticism creates a "self at peace," a self with power. The highest task of metaphysics (accomplished for example by Ibn Arabi, Boehme, Ramana Maharshi) is in a sense to self-destruct, to identify metaphysical & physical, transcendent & immanent, as ONE. Certain radical monists have pushed this doctrine far beyond mere pantheism or religious mysticism. An apprehension of the immanent oneness of being inspires certain antinomian heresies (the Ranters, the Assassins) whom we consider our ancestors.

Stirner himself seems deaf to the possible spiritual resonances of Individualism--& in this he belongs to the 19th century: born long after the deliquescence of Christendom, but long before the discovery of the Orient & of the hidden illuminist tradition in Western alchemy, revolutionary heresy & occult activism. Stirner quite correctly despised what he knew as "mysticism," a mere pietistic sentimentality based on self-abnegation & world hatred. Nietzsche nailed down the lid on "God" a few years later. Since then, who has dared to suggest that Individualism & mysticism might be reconciled & synthesized?

The missing ingredient in Stirner (Nietzsche comes closer) is a working concept of nonordinary consciousness. The realization of the unique self (or ubermensch) must reverberate & expand like waves or spirals or music to embrace direct experience or intuitive perception of the uniqueness of reality itself. This realization engulfs & erases all duality, dichotomy, & dialectic. It carries with itself, like an electric charge, an intense & wordless sense of value: it "divinizes" the self.

Being/consciousness/bliss (satchitananda) cannot be dismissed as merely another Stirnerian "spook" or "wheel in the head." It invokes no exclusively transcendent principle for which the Einzige must sacrifice his/her own-ness. It simply states that intense awareness of existence itself results in "bliss"--or in less loaded language, "valuative consciousness." The goal of the Unique after all is to possess everything; the radical monist attains this by identifying self with perception, like the Chinese inkbrush painter who "becomes the bamboo," so that "it paints itself."

Despite mysterious hints Stirner drops about a "union of Unique-ones" & despite Nietzsche's eternal "Yea" & exaltation of life, their Individualism seems somehow shaped by a certain coldness toward the other. In part they cultivated a bracing, cleansing chilliness against the warm suffocation of 19th century sentimentality & altruism; in part they simply despised what someone (Mencken?) called "Homo Boobensis."

And yet, reading behind & beneath the layer of ice, we uncover traces of a fiery doctrine--what Gaston Bachelard might have called "a Poetics of the Other." The Einzige's relation with the Other cannot be defined or limited by any institution or idea. And yet clearly, however paradoxically, the Unique depends for completeness on the Other, & cannot & will not be realized in any bitter isolation.

The examples of "wolf children" or enfants sauvages suggest that a human infant deprived of human company for too long will never attain conscious humanity--will never acquire language. The Wild Child perhaps provides a poetic metaphor for the Unique-one--and yet simultaneously marks the precise point where Unique & Other must meet, coalesce, unify--or else fail to attain & possess all of which they are capable.

The Other mirrors the Self--the Other is our witness. The Other completes the Self--the Other gives us the key to the perception of oneness-of-being. When we speak of being & consciousness, we point to the Self; when we speak of bliss we implicate the Other.

The acquisition of language falls under the sign of Eros-- all communication is essentially erotic, all relations are erotic. Avicenna & Dante claimed that love moves the very stars & planets in their courses--the Rg Veda & Hesiod's Theogony both proclaim Love the first god born after Chaos. Affections, affinities, aesthetic perceptions, beautiful creations, conviviali-ty--all the most precious possessions of the Unique-one arise from the conjunction of Self & Other in the constellation of Desire.

Here again the project begun by Individualism can be evolved & revivified by a graft with mysticism--specifically with tantra. As an esoteric technique divorced from orthodox Hinduism, tantra provides a symbolic framework ("Net of Jewels") for the identification of sexual pleasure & non-ordinary consciousness. All antinomian sects have contained some "tantrik" aspect, from the families of Love & Free Brethren & Adamites of Europe to the pederast sufis of Persia to the Taoist alchemists of China. Even classical anarchism has enjoyed its tantrik moments: Fourier's Phalansteries; the "Mystical Anarchism" of G. Ivanov & other fin-de-siÉcle Russian symbolists; the incestuous erotism of Arzibashaev's Sanine; the weird combination of Nihilism & Kali-worship which inspired the Bengali Terrorist Party (to which my tantrik guru Sri Kamanaransan Biswas had the honor of belonging)...

We, however, propose a much deeper syncretism of anarchy & tantra than any of these. In fact, we simply suggest that Individual Anarchism & Radical Monism are to be considered henceforth one and the same movement.

This hybrid has been called "spiritual materialism," a term which burns up all metaphysics in the fire of oneness of spirit & matter. We also like "Ontological Anarchy" because it suggests that being itself remains in a state of "divine Chaos," of all-potentiality, of continual creation.

In this flux only the jiva mukti, or "liberated individual," is self-realized, and thus monarch or owner of his perceptions and relations. In this ceaseless flow only desire offers any principle of order, and thus the only possible society (as Fourier understood) is that of lovers.

Anarchism is dead, long live anarchy! We no longer need the baggage of revolutionary masochism or idealist self-sacrifice--or the frigidity of Individualism with its disdain for conviviality, of living together--or the vulgar superstitions of 19th century atheism, scientism, and progressism. All that dead weight! Frowsy proletarian suitcases, heavy bourgeois steamer-trunks, boring philosophical portmanteaux--over the side with them!

We want from these systems only their vitality, their life-forces, daring, intransigence, anger, heedlessness--their power, their shakti. Before we jettison the rubbish and the carpetbags, we'll rifle the luggage for billfolds, revolvers, jewels, drugs and other useful items--keep what we like and trash the rest. Why not? Are we priests of a cult, to croon over relics and mumble our martyrologies?

Monarchism too has something we want--a grace, an ease, a pride, a superabundance. We'll take these, and dump the woes of authority & torture in history's garbage bin. Mysticism has something we need--"self-overcoming," exalted awareness, reservoirs of psychic potency. These we will expropriate in the name of our insurrection--and leave the woes of morality & religion to rot & decompose.

As the Ranters used to say when greeting any "fellow creature"--from king to cut-purse--"Rejoice! All is ours!"

INSTRUCTIONS FOR THE KALI YUGA

THE KALI YUGA STILL has 200,000 or so years to play--good news for advocates & avatars of CHAOS, bad news for Brahmins, Yahwists, bureaucrat-gods & their runningdogs.

I knew Darjeeling hid something for me soon as I heard the name--dorje ling--Thunderbolt City. In 1969 I arrived just before the monsoons. Old British hill station, summer hdqrs for Govt. of Bengal--streets in the form of winding wood staircases, the Mall with a View of Sikkim & Mt Katchenhunga--Tibetan temples & refugees--beautiful yellow-porcelain people called Lepchas (the real abo's)--Hindus, Moslems, Nepalese & Bhutanese Buddhists, & decaying Brits who lost their way home in '47, still running musty banks & tea-shoppes.

Met Ganesh Baba, fat white-bearded saddhu with overly-impeccable Oxford accent--never saw anyone smoke so much ganja, chillam after chillam full, then we'd wander the streets while he played ball with shrieking kids or picked fights in the bazaar, chasing after terrified clerks with his umbrella, then roaring with laughter.

He introduced me to Sri Kamanaransan Biswas, a tiny wispy middleage Bengali government clerk in a shabby suit, who offered to teach me Tantra. Mr Biswas lived in a tiny bungalow perched on a steep pine-tree misty hillside, where I visited him daily with pints of cheap brandy for puja & tippling--he encouraged me to smoke while we talked, since ganja too is sacred to Kali.

Mr Biswas in his wild youth was a member of the Bengali Terrorist Party, which included both Kali worshippers & heretic Moslem mystics as well as anarchists & extreme leftists. Ganesh Baba seemed to approve of this secret past, as if it were a sign of Mr Biswas's hidden tantrika strength, despite his outward seedy mild appearance.

We discussed my readings in Sir John Woodruffe ("Arthur Avalon") each afternoon, I walked there thru cold summer fogs, Tibetan spirit-traps flapping in the soaked breeze loomed out of the mist & cedars. We practiced the Tara-mantra and Tara-mudra (or Yoni-mudra), and studied the Tara-yantra diagram for magical purposes. Once we visited a temple to the Hindu Mars (like ours, both planet & war-god) where he bought a finger-ring made from an iron horseshoe nail & gave it to me. More brandy & ganja.

Tara: one of the forms of Kali, very similar in attributes: dwarfish, naked, four-armed with weapons, dancing on dead Shiva, necklace of skulls or severed heads, tongue dripping blood, skin a deep blue-grey the precise color of monsoon clouds. Every day more rain--mud-slides blocking roads. My Border Area Permit expires. Mr Biswas & I descend the slick wet Himalayas by jeep & train down to his ancestral city, Siliguri in the flat Bengali plains where the Ganges fingers into a sodden viridescent delta.

We visit his wife in the hospital. Last year a flood drowned Siliguri killing tens of thousands. Cholera broke out, the city's a wreck, algae-stained & ruined, the hospital's halls still caked with slime, blood, vomit, the liquids of death. She sits silent on her bed glaring unblinking at hideous fates. Dark side of the goddess. He gives me a colored lithograph of Tara which miraculously floated above the water & was saved.

That night we attend some ceremony at the local Kali-temple, a modest half-ruined little roadside shrine--torchlight the only illumination--chanting & drums with strange, almost African syncopation, totally unclassical, primordial & yet insanely complex. We drink, we smoke. Alone in the cemetery, next to a half-burnt corpse, I'm initiated into Tara Tantra. Next day, feverish & spaced-out, I say farewell & set out for Assam, to the great temple of Shakti's yoni in Gauhati, just in time for the annual festival. Assam is forbidden territory & I have no permit.

Midnight in Gauhati I sneak off the train, back down the tracks thru rain & mud up to my knees & total darkness, blunder at last into the city & find a bug-ridden hotel. Sick as a dog by this time. No sleep.

In the morning, bus up to the temple on a nearby mountain. Huge towers, pullulating deities, courtyards, outbuildings-- hundreds of thousands of pilgrims--weird saddhus down from their ice-caves squatting on tiger skins & chanting. Sheep & doves are being slaughtered by the thousands, a real hecatomb- -(not another white sahib in sight)--gutters running inch-deep in blood--curve-bladed Kali-swords chop chop chop, dead heads plocking onto the slippery cobblestones.

When Shiva chopped Shakti into 53 pieces & scattered them over the whole Ganges basin, her cunt fell here. Some friendly priests speak English & help me find the cave where Yoni's on display. By this time I know I'm seriously sick, but determined to finish the ritual. A herd of pilgrims (all at least one head shorter than me) literally engulfs me like an undertow-wave at the beach, & hurls me suspended down suffocating winding troglodyte stairs into claustrophobic womb-cave where I swirl nauseated & hallucinating toward a shapeless cone meteorite smeared in centuries of ghee & ochre. The herd parts for me, allows me to throw a garland of jasmine over the yoni.

A week later in Kathmandu I enter the German Missionary Hospital (for a month) with hepatitis. A small price to pay for all that knowledge--the liver of some retired colonel from a Kipling story!--but I know her, I know Kali. Yes absolutely the archetype of all that horror, yet for those who know, she becomes the generous mother. Later in a cave in the jungle above Rishikish I meditated on Tara for several days (with mantra, yantra, mudra, incense, & flowers) & returned to the serenity of Darjeeling, its beneficent visions.

Her age must contain horrors, for most of us cannot understand her or reach beyond the necklace of skulls to the garland of jasmine, knowing in what sense they are the same. To go thru CHAOS, to ride it like a tiger, to embrace it (even sexually) & absorb some of its shakti, its life-juice--this is the Path of Kali Yuga. Creative nihilism. For those who follow it she promises enlightenment & even wealth, a share of her temporal power.

The sexuality & violence serve as metaphors in a poem which acts directly on consciousness through the Image-ination--or else in the correct circumstances they can be openly deployed & enjoyed, embued with a sense of the holiness of every thing from ecstasy & wine to garbage & corpses.

Those who ignore her or see her outside themselves risk destruction. Those who worship her as ishta-devata, or divine self, taste her Age of Iron as if it were gold, knowing the alchemy of her presence.

AGAINST THE REPRODUCTION OF DEATH

ONE OF THE SIGNS of that End Time so many seem to anticipate would consist of a fascination with all the most negative & hateful detritus of that Time, a fascination felt by the very class of thinkers who consider themselves most perspicacious about the so-called apocalypse they warn us to beware. I'm speaking of people I know very well--those of the "spiritual right" (such as the neo-Guenonians with their obsession for signs of deca-dence)--& those of the post-philosophical left, the detached essayists of death, connoisseurs of the arts of mutilation.

For both these sets, all possible action in the world is smeared out onto one level plain--all become equally meaningless. For the Traditionalist, nothing matters but to prepare the soul for death (not only its own but the whole world's as well). For the "cultural critic" nothing matters but the game of identifying yet one more reason for despair, analyzing it, adding it to the catalogue.

Now the End of the World is an abstraction because it has never happened. It has no existence in the real world. It will cease to be an abstraction only when it happens--if it happens. (I do not claim to know "God's mind" on the subject--nor to possess any scientific knowledge about a still non-existent future). I see only a mental image & its emotional ramifications; as such I identify it as a kind of ghostly virus, a spook-sickness in myself which ought to be expunged rather than hypochondriacally coddled & indulged. I have come to despise the "End of the World" as an ideological icon held over my head by religion, state, & cultural milieu alike, as a reason for doing nothing.

I understand why the religious & political "powers" would want to keep me quaking in my shoes. Since only they offer even a chance of evading ragnarok (thru prayer, thru democracy, thru communism, etc.), I will sheepishly follow their dictates & dare nothing on my own. The case of the enlightened intellectuals, however, seems more puzzling at first. What power do they derive from this telling-the-beads of fear & gloom, sadism & hatred?

Essentially they gain smartness. Any attack on them must appear stupid, since they alone are clear-eyed enough to recognize the truth, they alone daring enough to show it forth in defiance of rude shit-kicking censors & liberal wimps. If I attack them as part of the very problem they claim to be discussing objectively, I will be seen as a bumpkin, a prude, a pollyanna. If I admit my hatred for the artifacts of their perception (books, artworks, performances) then I may be dismissed as merely squeamish (& so of course psychologically repressed), or else at the very least lacking in seriousness.

Many people assume that because I sometimes express myself as an anarchist boy-lover, I must also be "interested" in other ultra-postmodern ideas like serial child-murder, fascist ideology, or the photographs of Joel P. Witkin. They assume only two sides to any issue--the hip side & the unhip side. A marxist who objected to all this death-cultishness as anti-progressive would be thought as foolish as a Xtian fundamentalist who believed it immoral.

I maintain that (as usual) many sides exist to this issue rather than only two. Two-sided issues (creationism vs darwinism, "choice" vs "pro-life," etc.) are all without exception delusions, spectacular lies.

My position is this: I am all too well aware of the "intelligence" which prevents action. I myself possess it in abundance. Every once in a while however I have managed to behave as if I were stupid enough to try to change my life. Sometimes I've used dangerous stupifiants like religion, marijuana, chaos, the love of boys. On a few occasions I have attained some degree of success--& I say this not to boast but rather to bear witness. By overthrowing the inner icons of the End of the World & the Futility of all mundane endeavor, I have (rarely) broken through into a state which (by comparison with all I'd known) appeared to be one of health. The images of death & mutilation which fascinate our artists & intellectuals appear to me--in the remembered light of these experiences--tragically inappropriate to the real potential of existence & of discourse about exis-tence.

Existence itself may be considered an abyss possessed of no meaning. I do not read this as a pessimistic statement. If it be true, then I can see in it nothing else but a declaration of autonomy for my imagination & will--& for the most beautiful act they can conceive with which to bestow meaning upon existence.

Why should I emblemize this freedom with an act such as murder (as did the existentialists) or with any of the ghoulish tastes of the eighties? Death can only kill me once--till then I am free to express & experience (as much as I can) a life & an art of life based on self-valuating "peak experiences," as well as "conviviality" (which also possesses its own reward).

The obsessive replication of Death-imagery (& its reproduction or even commodification) gets in the way of this project just as obstructively as censorship or media-brainwashing. It sets up negative feedback loops--it is bad juju. It helps no one conquer fear of death, but merely inculcates a morbid fear in place of the healthy fear all sentient creatures feel at the smell of their own mortality.

This is not to absolve the world of its ugliness, or to deny that truly fearful things exist in it. But some of these things can be overcome--on the condition that we build an aesthetic on the overcoming rather than the fear.

I recently attended a gay dance/poetry performance of uncompromising hipness: the one black dancer in the troupe had to pretend to fuck a dead sheep.

Part of my self-induced stupidity, I confess, is to believe (& even feel) that art can change me, & change others. That's why I write pornography & propaganda--to cause change. Art can

never mean as much as a love affair, perhaps, or an insurrection. But...to a certain extent...it works.

Even if I'd given up all hope in art, however, all expectation of
exaltation, I would still refuse to put up with art that merely
exacerbates my misery, or indulges in schadenfreude, "delight
in the misery of others." I turn away from certain art as a dog
would turn away howling from the corpse of its companion. I'd
like to renounce the sophistication which would permit me to
sniff it with detached curiosity as yet another example of post-
industrial decomposition.

Only the dead are truly smart, truly cool. Nothing touches them.
While I live, however, I side with bumbling suffering crooked
life, with anger rather than boredom, with sweet lust, hunger &
carelessness...against the icy avant-guard & its fashionable
premonitions of the sepulcher.

RINGING DENUNCIATION OF SURREALISM

(For Harry Smith)

AT THE SURREALIST FILM show, someone asked Stan Brakhage
about the media's use of surrealism (MTV, etc.); he answered
that it was a "damn shame." Well, maybe it is & maybe it isn't
(does popular kultur ipso facto lack all inspiration?)--but
granting that on some level the media's appropriation of
surrealism is a damn shame, are we to believe that there was
nothing in surrealism that allowed this theft to occur?

The return of the repressed means the return of the paleolithic--
not a return to the Old Stone Age, but a spiralling around on a
new level of the gyre. (After all, 99.9999% of human experience
is of hunting/gathering, with agriculture & industry a mere oil

slick on the deep well of non-history.) Paleolithic equals pre-Work ("original leisure society"). Post-Work (Zerowork) equals "Psychic Paleolithism."

All projects for the "liberation of desire" (Surrealism) which remain enmeshed in the matrix of Work can only lead to the commodification of desire. The Neolithic begins with desire for commodities (agricultural surplus), moves on to the production of desire (industry), & ends with the implosion of desire (advertising). The Surrealist liberation of desire, for all its aesthetic accomplishments, remains no more than a subset of production--hence the wholesaling of Surrealism to the Communist Party & its Work-ist ideology (not to mention attendant misogyny & homophobia). Modern leisure, in turn, is simply a subset of Work (hence its commodification)--so it is no accident that when Surrealism closed up shop, the only customers at the garage sale were ad execs.

Advertising, using Surrealism's colonization of the unconscious to create desire, leads to the final implosion of Surrealism. It's not just a "damn shame & a disgrace," not a simple appropriation. Surrealism was made for advertising, for commodification. Surrealism is in fact a betrayal of desire.

And yet, out of this abyss of meaning, desire still rises, innocent as a new-hatched phoenix. Early Berlin dada (which rejected the return of the art-object) for all its faults provides a better model for dealing with the implosion of the social than Surrealism could ever do--an anarchist model, or perhaps (in anthro-jargon) a non-authoritarian model, a destruction of all ideology, of all chains of law. As the structure of Work/Leisure crumbles into emptiness, as all forms of control vanish in the dissolution of meaning, the Neolithic seems bound to vanish as well, with all its temples & granaries & police, to be replaced by some return of hunting/gathering on the psychic level--a re-nomadization. Everything's imploding & disappearing--the oedipal family, education, even the unconscious itself (as Andr-Codrescu says). Let's not mistake this for Armageddon (let's resist the seduction of apocalypse, the eschatological con)--it's not the world coming to an end--only the empty husks of the social, catching fire & disappearing.

Surrealism must be junked along with all the other beautiful bric-a-brac of agricultural priestcraft & vapid control-systems. No one knows what's coming, what misery, what spirit of wildness, what joy--but the last thing we need on our voyage is another set of commissars--popes of our dreams--daddies. Down with Surrealism...

--Naropa, July 9, 1988

FOR A CONGRESS OF WEIRD RELIGIONS

WE'VE LEARNED TO DISTRUST the verb to be, the word is--let's say rather: note the striking resemblance between the concept SATORI & the concept REVOLUTION OF EVERYDAY LIFE--in both cases: a perception of the "ordinary" with extraordinary consequences for consciousness & action. We can't use the phrase "is like" because both concepts (like all concepts, all words for that matter) come crusted with accretions--each burdened with all its psycho-cultural baggage, like guests who arrive suspiciously overly well-supplied for the weekend.

So allow me the old-fashioned Beat-Zennish use of satori, while simultaneously emphasizing--in the case of the Situationist slogan--that one of the roots of its dialectic can be traced to dada & Surrealism's notion of the "marvelous" erupting from (or into) a life which only seems suffocated by the banal, by the miseries of abstraction & alienation. I define my terms by making them more vague, precisely in order to avoid the orthodoxies of both Buddhism & Situationism, to evade their ideologico-semantic traps--those broken-down language machines! Rather, I propose we ravage them for parts, an act of cultural bricolage. "Revolution" means just another turn of the crank--while religious orthodoxy of any sort leads logically to a veritable government of cranks. Let's not idolize satori by imagining it the monopoly of mystic monks, or as contingent on any moral code; & rather than fetishize the Leftism of '68 we prefer Stirner's term "insurrection" or "uprising," which escapes the built-in implications of a mere change of authority.

This constellation of concepts involves "breaking rules" of ordered perception to arrive at direct experiencing, somewhat analogous to the process whereby chaos spontaneously resolves into fractal nonlinear orders, or the way in which "wild" creative energy resolves as play & poesis. "Spontaneous order" out of "chaos" in turn evokes the anarchist Taoism of the Chuang Tzu. Zen may be accused of lacking awareness of the "revolutionary" implications of satori, while the Situationists can be criticized for ignoring a certain "spirituality" inherent in the self-realization & conviviality their cause demands. By identifying satori with the r. of e.d.l. we're performing a bit of a shotgun marriage fully as remarkable as the Surrealists' famous mating of an umbrella & sewing machine or whatever it was. Miscegenation. The race-mixing advocated by Nietzsche, who was attracted, no doubt, by the sexiness of the half-caste.

I'm tempted to try to describe the way satori "is" like the r. of e.d.l.--but I can't. Or to put it another way: nearly all I write revolves around this theme; I would have to repeat nearly everything in order to elucidate this single point. Instead, as an appendix, I offer one more curious coincidence or interpenetration of 2 terms, one from Situationism again & the other this time from sufism. The d-rive or "drift" was conceived as an exercise in deliberate revolutionizing of everyday life--a sort of aimless wandering thru city streets, a visionary urban nomadism involving an openness to "culture as nature" (if I grasp the idea correctly)--which by its sheer duration would inculcate in the drifters a propensity to experience the marvelous; not always in its beneficent form perhaps, but hopefully always productive of insight--whether thru architecture, the erotic, adventure, drink & drugs, danger, inspiration, whatever--into the intensity of unmediated perception & experience.

The parallel term in sufism would be "journeying to the far horizons" or simply "journeying," a spiritual exercise which combines the urban & nomadic energies of Islam into a single trajectory, sometimes called "the Caravan of Summer." The dervish vows to travel at a certain velocity, perhaps spending no more than 7 nights or 40 nights in one city, accepting whatever comes, moving wherever signs & coincidences or simply whims may lead, heading from power-spot to power-spot, conscious of "sacred geography," of itinerary as meaning, of topology as symbology. Here's another constellation: Ibn Khaldun, On the Road (both Jack Kerouac's & Jack London's), the form of the picaresque novel in general, Baron Munchausen, wanderjahr, Marco Polo, boys in a suburban summer forest, Arthurian knights out questing for trouble, queers out cruising for boys, pub-crawling with Melville, Poe, Baudelaire--or canoeing with Thoreau in Maine...travel as the antithesis of tourism, space rather than time. Art project: the construction of a "map" bearing a 1:1 ratio to the "territory" explored. Political project: the construction of shifting "autonomous zones" within an invisible nomadic network (like the Rainbow Gatherings). Spiritual project: the creation or discovery of pilgrimages in which the concept "shrine" has been replaced (or esotericized) by the concept "peak experience."

What I'm trying to do here (as usual) is to provide a sound irrational basis, a strange philosophy if you like, for what I call the Free Religions, including the Psychedelic & Discordian currents, non-hierarchical neo-paganism, antinomian heresies, chaos & Kaos Magik, revolutionary HooDoo, "unchurched" & anarchist Christians, Magical Judaism, the Moorish Orthodox Church, Church of the SubGenius, the Faeries, radical Taoists, beer mystics, people of the Herb, etc., etc.

Contrary to the expectations of 19th century radicals, religion has not gone away--perhaps we'd be better off if it had--but has instead increased in power, seemingly in proportion to the global increase in the realm of technology & rational control. Both fundamentalism & the New Age derive some force from deep & widespread dissatisfaction with the System that works against all perception of the marvelousness of everyday life--call it Babylon or the Spectacle, Capital or Empire, Society of Simulation or of soulless mechanism--what you wish. But these two religious forces divert the very desire for the authentic toward overpowering & oppressive new abstractions (morality in the case of fundamentalism, commodification in the case of the New Age), & for this reason can quite properly be called "reactionary."

Just as cultural radicals will seek to infiltrate & subvert the popular media, & just as political radicals will perform similar functions in the spheres of Work, Family, & other social organizations, so there exists a need for radicals to penetrate the institution of religion itself rather than merely continue to mouth 19th century platitudes about atheistic materialism. It's going to happen anyway--better to approach it with consciousness, with grace & style.

Having once lived near the Hdqrs of the World Council of Churches, I like the possibility of a Free Churches parody version--parody being one of our chief strategies (or call it d-tournement or deconstruction or creative destruction)--a sort of loose network (I dislike that word; let's call it a "webwork" instead) of weird cults & individuals providing conversation & services for each other, out of which might begin to emerge a trend or tendency or "current" (in magical terms) strong enough to wreak some psychic havoc on the Fundies & New Agers, even the ayatollahs & the Papacy, convivial enough for us to disagree with each other & yet still give great parties--or conclaves, or ecumenical councils, or World Congresses--which we anticipate with glee.

The Free Religions may offer some of the only possible spiritual alternatives to televangelist stormtroopers & pinhead crystal-channelers (not to mention the established religions), & will thus become more & more important, more & more vital in a future where the demand for the eruption of the marvelous into the ordinary will become the most ringing, poignant & tumultuous of all political demands--a future which will begin (wait a minute, lemme check my clock)...7, 6, 5, 4, 3, 2, 1...NOW.

HOLLOW EARTH

SUBTERRANEAN REGIONS OF THE continent excavated in cyclopaean caverns, cathedralspace fractal networks, labyrin-thine gargantuan tunnels, slow black underground rivers, unmoving stygian lakes, pure & slightly luminiferous, slim waterfalls plunging down watersmooth rock, cataracting round petrified forests of stalactites & stalagmites in spelunker-bewildering blind-fish complexity & unfathomable vast-ness...Who dug this hollow earth beneath the ice foreseen by Poe, by certain paranoid German occultists, Shaverian UFO freaks? Was Earth once colonized in the time of Gondwana or MU by some Elder Race? their reptilian skeletons still mouldering in the farthest secret mazes of the cavern system? Sluggish backwaters, dead-end canals, stagnant pools far from the centers of civilization like Little America, Transport City, or Nan Chi Han, down in the dark recesses and boondocks of the Antarctic caves, fungus & albino fern. We suspect them of mutations, amphibian webbed fingers and toes, degenerate habits--Kallikaks of the Hollow Earth, Lovecraftian renegades, hermits, skulking incestuous smugglers, runaway criminals, anarchists forced into hiding after the Entropy Wars, fugitives from Genetic Puritanism, dissident Chinese Tongs & Yellow Turban fanatics, lascar cave-pirates, pale shiftless whitetrash from the prolewarrens of the industrial domes along Thwait's Tongue & the Walgreen Coast & Edsel-Ford-Land--the Trogs have kept alive for over 200 years the folk-memory of the Autonomous Zone, the myth that someday it will appear again...Taoism, libertine philosophy, Indonesian sorcery, cult of the Cave Mother (or Mothers), identified by some scholars with the Javanese sea/moon goddess Loro Kidul, by others with a minor deity of the South Pole Star Sect, the "Jade Goddess"...manuscripts (written in Bahasa Ingliss the pidgin dialect of the deep caves) contain mangled quotations from Nietzsche & Chuang Tzu...Trade consists of occasional precious gems and cultivation of white poppy, fungus, over a dozen different species of "magic" mushrooms...Shallow Lake Erebus, 5 miles across, dotted with stalagmitic islets choked with fern & kudzu & black dwarf pine, held in a cave so vast it sometimes creates its own weather...The town belongs officially to Little America but most of the inhabitants are Trogs living off the Shiftless Dole--& the deep-cave tribal country lies just across the Lake. Riffraff, artists, drug addicts, sorcerers, smugglers, remittance-men & perverts live in crumbling basalt-&-synthplast hotels half-encrusted with pale green vines, along the lakefront, an avenue of squalid cafes, gem emporia guarded by armed ninjas, chinese krill-noodle shops, the crystal-tinselled hall for slow fusion-gamelan dancers, boys practicing their mudras on sleepy electronic dark blue afternoons to the rippling of synthgongs and metallophones...& below the pier perhaps a few desultory bathers along the black beach, genuine low-budget tourists gawking at the shrine behind the bazaar where pallid old Trog pamongs tranced out on fungus drool & roll up their eyes, breathe in the fumes of heavy incense, everything seems suddenly menacingly bright, flickering with significance...a few cases of webbed fingers but the rumors of ritual promiscuity are true enough. I was living in a Trog fishing village across the lake from Erebus in a rented room above the baitshop...rural sloth & degenerate superstitious rites of sensual abandon, the larval & unhealthy mysteries of the chthonic mutant downtrodden Trogs, lazy shiftless no-count hicks...Little America, so christian & free of mutation, eugenic & orderly, where everyone lives jacked into the fleshless realm of ancient software & holography, so euclidean, newtonian, clean & patriotic--L.A. will never understand this innocent filth-sorcery, this "spiritual material-ism," this slavery to the volcanic desires of secret cave-boy gangs like laughing flowers jetting with dynamo erections pulsing up pure life curved taut as bows, & the smell of water, pond-scum, nightblooming white flowers, jasmine & datura, urine, children's wet hair, sperm & mud...possessed by cave-spirits, perhaps ghosts of ancient aliens now wandering as demons seeking to renew long-lost pleasures of flesh & substance. Or else the Zone has already been reborn, already a nexus of autonomy, a spreading virus of chaos in its most exuberant clandestine form, white toadstools springing up on the spots where Trog boys have masturbated alone in the dark...

NIETZSCHE & THE DERVISHES

RENDAN, "THE CLEVER ONES." The sufis use a technical term rend (adj. rendi, pl. rendan) to designate one "clever enough to drink wine in secret without getting caught": the dervish version of "Permissible Dissimulation" (taqiyya, whereby Shiites are permitted to lie about their true affiliation to avoid persecution as well as advance the purpose of their propaganda).

On the plane of the "Path," the rend conceals his spiritual state (hal) in order to contain it, work on it alchemically, enhance it. This "cleverness" explains much of the secrecy of the Orders, altho it remains true that many dervishes do literally break the rules of Islam (shariah), offend tradition (sunnah), and flout the customs of their society--all of which gives them reason for real secrecy.

Ignoring the case of the "criminal" who uses sufism as a mask-- or rather not sufism per se but dervishism, almost a synonym in Persia for laid-back manners & by extension a social laxness, a style of genial and poor but elegant amorality--the above definition can still be considered in a literal as well as metaphor- ical sense. That is: some sufis do break the Law while still allowing that the Law exists & will continue to exist; & they do so from spiritual motives, as an exercise of will (himmah).

Nietzsche says somewhere that the free spirit will not agitate for the rules to be dropped or even reformed, since it is only by breaking the rules that he realizes his will to power. One must prove (to oneself if no one else) an ability to overcome the rules of the herd, to make one's own law & yet not fall prey to the rancor & resentment of inferior souls which define law & custom in ANY society. One needs, in effect, an individual equivalent of war in order to achieve the becoming of the free spirit--one needs an inert stupidity against which to measure one's own movement & intelligence.

Anarchists sometimes posit an ideal society without law. The few anarchist experiments which succeeded briefly (the

Makhnovists, Catalan) failed to survive the conditions of war which permitted their existence in the first place--so we have no way of knowing empirically if such an experiment could outlive the onset of peace.

Some anarchists, however, like our late friend the Italian Stirnerite "Brand," took part in all sorts of uprisings and revolutions, even communist and socialist ones, because they found in the moment of insurrection itself the kind of freedom they sought. Thus while utopianism has so far always failed, the individualist or existentialist anarchists have succeeded inasmuch as they have attained (however briefly) the realization of their will to power in war.

Nietzsche's animadversions against "anarchists" are always aimed at the egalitarian-communist narodnik martyr types, whose idealism he saw as yet one more survival of post-Xtian moralism--altho he sometimes praises them for at least having the courage to revolt against majoritarian authority. He never mentions Stirner, but I believe he would have classified the Individualist rebel with the higher types of "criminals," who represented for him (as for Dostoyevsky) humans far superior to the herd, even if tragically flawed by their obsessiveness and perhaps hidden motivations of revenge.

The Nietzschean overman, if he existed, would have to share to some degree in this "criminality" even if he had overcome all obsessions and compulsions, if only because his law could never agree with the law of the masses, of state & society. His need for "war" (whether literal or metaphorical) might even persuade him to take part in revolt, whether it assumed the form of insurrection or only of a proud bohemianism.

For him a "society without law" might have value only so long as it could measure its own freedom against the subjection of others, against their jealousy & hatred. The lawless & short-lived "pirate utopias" of Madagascar & the Caribbean, D'Annunzio's Republic of Fiume, the Ukraine or Barcelona--these would attract him because they promised the turmoil of becoming & even "failure" rather than the bucolic somnolence of a "perfected" (& hence dead) anarchist society.

In the absence of such opportunities, this free spirit would disdain wasting time on agitation for reform, on protest, on visionary dreaming, on all kinds of "revolutionary martyrdom"--in short, on most contemporary anarchist activity. To be rendi, to drink wine in secret & not get caught, to accept the rules in order to break them & thus attain the spiritual lift or energy-rush of danger & adventure, the private epiphany of overcoming all interior police while tricking all outward authority--this might be a goal worthy of such a spirit, & this might be his definition of crime.

(Incidentally, I think this reading helps explain N's insistence on the MASK, on the secretive nature of the proto-overman, which disturbs even intelligent but somewhat liberal commentators like Kaufman. Artists, for all that N loves them, are criticized for telling secrets. Perhaps he failed to consider that--paraphrasing A. Ginsberg--this is our way of becoming "great"; and also that--paraphrasing Yeats--even the truest secret becomes yet another mask.)

As for the anarchist movement today: would we like just once to stand on ground where laws are abolished & the last priest is strung up with the guts of the last bureaucrat? Yeah sure. But we're not holding our breath. There are certain causes (to quote the Neech again) that one fails to quite abandon, if only because of the sheer insipidity of all their enemies. Oscar Wilde might have said that one cannot be a gentleman without being something of an anarchist--a necessary paradox, like N's "radical aristocratism."

This is not just a matter of spiritual dandyism, but also of existential commitment to an underlying spontaneity, to a philosophical "tao." For all its waste of energy, in its very formlessness, anarchism alone of all the ISMs approaches that one type of form which alone can interest us today, that strange attractor, the shape of chaos--which (one last quote) one must have within oneself, if one is to give birth to a dancing star.

--Spring Equinox, 1989

RESOLUTION FOR THE 1990's: BOYCOTT COP CULTURE!!!

IF ONE FICTIONAL FIGURE can be said to have dominated the popcult of the eighties, it was the Cop. Fuckin' police every-where you turned, worse than real life. What an incredible bore.

Powerful Cops--protecting the meek and humble--at the expense of a half-dozen or so articles of the Bill of Rights--"Dirty Harry." Nice human cops, coping with human perversity, coming out sweet 'n' sour, you know, gruff & knowing but still soft inside--Hill Street Blues--most evil TV show ever. Wiseass black cops scoring witty racist remarks against hick white cops, who nevertheless come to love each other--Eddie Murphy, Class Traitor. For that masochist thrill we got wicked bent cops who threaten to topple our Kozy Konsensus Reality from within like Giger-designed tapeworms, but naturally get blown away just in the nick of time by the Last Honest Cop, Robocop, ideal amalgam of prosthesis and sentimentality.

We've been obsessed with cops since the beginning--but the rozzers of yore played bumbling fools, Keystone Kops, Car 54 Where Are You, booby-bobbies set up for Fatty Arbuckle or Buster Keaton to squash & deflate. But in the ideal drama of the eighties, the "little man" who once scattered bluebottles by the hundred with that anarchist's bomb, innocently used to light a cigarette--the Tramp, the victim with the sudden power of the pure heart--no longer has a place at the center of narrative. Once "we" were that hobo, that quasi-surrealist chaote hero who wins thru wu-wei over the ludicrous minions of a despised & irrelevant Order. But now "we" are reduced to the status of victims without power, or else criminals. "We" no longer occupy that central role; no longer the heros of our own stories, we've been marginalized & replaced by the Other, the Cop.

Thus the Cop Show has only three characters--victim, criminal, and policeperson--but the first two fail to be fully human--only the pig is real. Oddly enough, human society in the eighties (as seen in the other media) sometimes appeared to consist of the same three cliche/archetypes. First the victims, the whining minorities bitching about "rights"--and who pray tell did not belong to a "minority" in the eighties? Shit, even cops complained about their "rights" being abused. Then the criminals: largely non-white (despite the obligatory & hallucinatory "integration" of the media), largely poor (or else obscenely rich, hence even more alien), largely perverse (i.e. the forbidden mirrors of "our" desires). I've heard that one out of four households in America is robbed every year, & that every year nearly half a million of us are arrested just for smoking pot. In the face of such statistics (even assuming they're "damned lies") one wonders who is NOT either victim or criminal in our police-state-of-consciousness. The fuzz must mediate for all of us, however fuzzy the interface-- they're only warrior-priests, however profane. America's Most Wanted--the most successful TV game show of the eighties--opened up for all of us the role of Amateur Cop, hitherto merely a media fantasy of middleclass resentment & revenge. Naturally the truelife Cop hates no one so much as the vigilante--look what happens to poor &/or non-white neighborhood self-protection groups like the Muslims who tried to eliminate crack dealing in Brooklyn: the cops busted the Muslims, the pushers went free. Real vigilantes threaten the monopoly of enforcement, lÉse majest-, more abominable than incest or murder. But media(ted) vigilantes function perfectly within the CopState; in fact, it would be more accurate to think of them as unpaid (not even a set of matched luggage!) informers: telemetric snitches, electro-stoolies, ratfinks-for-a-day.

What is it that "America most wants"? Does this phrase refer to criminals--or to crimes, to objects of desire in their real presence, unrepresented, unmediated, literally stolen & appropriated? America most wants...to fuck off work, ditch the spouse, do drugs (because only drugs make you feel as good as the people in TV ads appear to be), have sex with nubile jailbait, sodomy, burglary, hell yes. What unmediated pleasures are NOT illegal? Even outdoor barbecues violate smoke ordinances nowadays. The simplest enjoyments turn us against some law; finally pleasure becomes too stress-inducing, and only TV remains--and the pleasure of revenge, vicarious betrayal, the sick thrill of the tattletale. America can't have what it most wants, so it has America's Most Wanted instead. A nation of schoolyard toadies sucking up to an elite of schoolyard bullies.

Of course the program still suffers from a few strange reality-glitches: for example, the dramatized segments are enacted cinema verit-style by actors; some viewers are so stupid they believe they're seeing actual footage of real crimes. Hence the actors are being continually harassed & even arrested, along with (or instead of) the real criminals whose mugshots are flashed after each little documentoid. How quaint, eh? No one really experiences anything--everyone reduced to the status of ghosts--media-images break off & float away from any contact with actual everyday life--PhoneSex--CyberSex. Final transcendence of the body: cybergnosis.

The media cops, like televangelical forerunners, prepare us for the advent, final coming or Rapture of the police state: the "Wars" on sex and drugs: total control totally leached of all content; a map with no coordinates in any known space; far beyond mere Spectacle; sheer ecstasy ("standing-outside-the-body"); obscene simulacrum; meaningless violent spasms elevated to the last principle of governance. Image of a country consumed by images of self-hatred, war between the schizoid halves of a split personality, Super-Ego vs the Id Kid, for the heavyweight championship of an abandoned landscape, burnt, polluted, empty, desolate, unreal. Just as the murder-mystery is always an exercise in sadism, so the cop-fiction always involves the contemplation of control. The image of the inspector or detective measures the image of "our" lack of autonomous substance, our transparency before the gaze of authority. Our perversity, our helplessness. Whether we imagine them as "good" or "evil," our obsessive invocation of the eidolons of the Cops reveals the extent to which we have accepted the manichaean worldview they symbolize. Millions of tiny cops swarm everywhere, like the qlippoth, larval hungry ghosts--they fill the screen, as in Keaton's famous two-reeler, overwhelming the foreground, an Antarctic where nothing moves but hordes of sinister blue penguins.

We propose an esoteric hermeneutical exegesis of the Surrealist slogan "Mort aux vaches!" We take it to refer not to the deaths of individual cops ("cows" in the argot of the period)--mere leftist revenge fantasy--petty reverse sadism--but rather to the death of the image of the flic, the inner Control & its myriad reflections in the NoPlace Place of the media--the "gray room" as Burroughs calls it. Self-censorship, fear of one's own desires, "conscience" as the interiorized voice of consensus-authority. To assassinate these "security forces" would indeed release floods of libidinal energy, but not the violent running-amok predicted by the theory of Law 'n' Order.

Nietzschean "self-overcoming" provides the principle of organization for the free spirit (as also for anarchist society, at least in theory). In the police-state personality, libidinal energy is dammed & diverted toward self-repression; any threat to Control results in spasms of violence. In the free-spirit personality, energy flows unimpeded & therefore turbulently but gently--its chaos finds its strange attractor, allowing new spontaneous orders to emerge.

In this sense, then, we call for a boycott of the image of the Cop, & a moratorium on its production in art. In this sense...

MORT AUX VACHES!

THE TEMPORARY AUTONOMOUS ZONE

"...this time however I come as the victorious Dionysus, who will turn the world into a holiday...Not that I have much time..."

--Nietzsche (from his last "insane" letter to Cosima Wagner)

Pirate Utopias

THE SEA-ROVERS AND CORSAIRS of the 18th century created an "information network" that spanned the globe: primitive and devoted primarily to grim business, the net nevertheless functioned admirably. Scattered throughout the net were islands, remote hideouts where ships could be watered and provisioned, booty traded for luxuries and necessities. Some of these islands supported "intentional communities," whole mini-societies living consciously outside the law and determined to keep it up, even if only for a short but merry life.

Some years ago I looked through a lot of secondary material on piracy hoping to find a study of these enclaves--but it appeared as if no historian has yet found them worthy of analysis. (William Burroughs has mentioned the subject, as did the late British anarchist Larry Law--but no systematic research has been carried out.) I retreated to primary sources and constructed my own theory, some aspects of which will be discussed in this essay. I called the settlements "Pirate Utopias."

Recently Bruce Sterling, one of the leading exponents of Cyberpunk science fiction, published a near-future romance based on the assumption that the decay of political systems will lead to a decentralized proliferation of experiments in living: giant worker-owned corporations, independent enclaves devoted to "data piracy," Green-Social-Democrat enclaves, Zerowork enclaves, anarchist liberated zones, etc. The information economy which supports this diversity is called the Net; the enclaves (and the book's title) are Islands in the Net.

The medieval Assassins founded a "State" which consisted of a network of remote mountain valleys and castles, separated by thousands of miles, strategically invulnerable to invasion, connected by the information flow of secret agents, at war with all governments, and devoted only to knowledge. Modern technology, culminating in the spy satellite, makes this kind of autonomy a romantic dream. No more pirate islands! In the future the same technology--freed from all political control-- could make possible an entire world of autonomous zones. But for now the concept remains precisely science fiction--pure speculation.

Are we who live in the present doomed never to experience autonomy, never to stand for one moment on a bit of land ruled only by freedom? Are we reduced either to nostalgia for the past or nostalgia for the future? Must we wait until the entire world is freed of political control before even one of us can claim to know freedom? Logic and emotion unite to condemn such a supposition. Reason demands that one cannot struggle for what one does not know; and the heart revolts at a universe so cruel as to visit such injustices on our generation alone of humankind.

To say that "I will not be free till all humans (or all sentient creatures) are free" is simply to cave in to a kind of nirvana-stupor, to abdicate our humanity, to define ourselves as losers.

I believe that by extrapolating from past and future stories about "islands in the net" we may collect evidence to suggest that a certain kind of "free enclave" is not only possible in our time but also existent. All my research and speculation has crystallized around the concept of the TEMPORARY AUTO-NOMOUS ZONE (hereafter abbreviated TAZ). Despite its synthesizing force for my own thinking, however, I don't intend the TAZ to be taken as more than an essay ("attempt"), a suggestion, almost a poetic fancy. Despite the occasional Ranterish enthusiasm of my language I am not trying to construct political dogma. In fact I have deliberately refrained from defining the TAZ--I circle around the subject, firing off exploratory beams. In the end the TAZ is almost self-explanatory. If the phrase became current it would be understood without difficulty...understood in action.

Waiting for the Revolution

HOW IS IT THAT "the world turned upside-down" always manages to Right itself? Why does reaction always follow revolution, like seasons in Hell?

Uprising, or the Latin form insurrection, are words used by historians to label failed revolutions--movements which do not match the expected curve, the consensus-approved trajectory: revolution, reaction, betrayal, the founding of a stronger and even more oppressive State--the turning of the wheel, the return of history again and again to its highest form: jackboot on the face of humanity forever.

By failing to follow this curve, the up-rising suggests the possibility of a movement outside and beyond the Hegelian spiral of that "progress" which is secretly nothing more than a vicious circle. Surgo--rise up, surge. Insurgo--rise up, raise oneself up. A bootstrap operation. A goodbye to that wretched parody of the karmic round, historical revolutionary futility. The slogan "Revolution!" has mutated from tocsin to toxin, a malign pseudo-Gnostic fate-trap, a nightmare where no matter how we struggle we never escape that evil Aeon, that incubus the State, one State after another, every "heaven" ruled by yet one more evil angel.

If History IS "Time," as it claims to be, then the uprising is a moment that springs up and out of Time, violates the "law" of History. If the State IS History, as it claims to be, then the insurrection is the forbidden moment, an unforgivable denial of the dialectic--shimmying up the pole and out of the smokehole, a shaman's maneuver carried out at an "impossible angle" to the universe. History says the Revolution attains "permanence," or at least duration, while the uprising is "temporary." In this sense an uprising is like a "peak experience" as opposed to the standard of "ordinary" consciousness and experience. Like festivals, uprisings cannot happen every day--otherwise they would not be "nonordinary." But such moments of intensity give shape and meaning to the entirety of a life. The shaman returns--you can't stay up on the roof forever--but things have changed, shifts and integrations have occurred--a difference is made.

You will argue that this is a counsel of despair. What of the anarchist dream, the Stateless state, the Commune, the autonomous zone with duration, a free society, a free culture? Are we to abandon that hope in return for some existentialist acte gratuit? The point is not to change consciousness but to change the world.

I accept this as a fair criticism. I'd make two rejoinders never-theless; first, revolution has never yet resulted in achieving this dream. The vision comes to life in the moment of uprising--but as soon as "the Revolution" triumphs and the State returns, the dream and the ideal are already betrayed. I have not given up hope or even expectation of change--but I distrust the word Revolution. Second, even if we replace the revolutionary approach with a concept of insurrection blossoming sponta- neously into anarchist culture, our own particular historical situation is not propitious for such a vast undertaking. Absolute- ly nothing but a futile martyrdom could possibly result now from a head-on collision with the terminal State, the megacor- porate information State, the empire of Spectacle and Simula- tion. Its guns are all pointed at us, while our meager weaponry finds nothing to aim at but a hysteresis, a rigid vacuity, a Spook capable of smothering every spark in an ectoplasm of informa- tion, a society of capitulation ruled by the image of the Cop and the absorbant eye of the TV screen.

In short, we're not touting the TAZ as an exclusive end in itself, replacing all other forms of organization, tactics, and goals. We recommend it because it can provide the quality of enhance- ment associated with the uprising without necessarily leading to violence and martyrdom. The TAZ is like an uprising which does not engage directly with the State, a guerilla operation which liberates an area (of land, of time, of imagination) and then dissolves itself to re-form elsewhere/elsewhen, before the State can crush it. Because the State is concerned primarily with Simulation rather than substance, the TAZ can "occupy" these areas clandestinely and carry on its festal purposes for quite a while in relative peace. Perhaps certain small TAZs have lasted whole lifetimes because they went unnoticed, like hillbilly enclaves--because they never intersected with the Spectacle, never appeared outside that real life which is invisible to the agents of Simulation.

Babylon takes its abstractions for realities; precisely within this margin of error the TAZ can come into existence. Getting the

TAZ started may involve tactics of violence and defense, but its greatest strength lies in its invisibility--the State cannot recognize it because History has no definition of it. As soon as the TAZ is named (represented, mediated), it must vanish, it will vanish, leaving behind it an empty husk, only to spring up again somewhere else, once again invisible because undefinable in terms of the Spectacle. The TAZ is thus a perfect tactic for an era in which the State is omnipresent and all-powerful and yet simultaneously riddled with cracks and vacancies. And because the TAZ is a microcosm of that "anarchist dream" of a free culture, I can think of no better tactic by which to work toward that goal while at the same time experiencing some of its benefits here and now.

In sum, realism demands not only that we give up waiting for "the Revolution" but also that we give up wanting it. "Uprising," yes--as often as possible and even at the risk of violence. The spasming of the Simulated State will be "spectacular," but in most cases the best and most radical tactic will be to refuse to engage in spectacular violence, to withdraw from the area of simulation, to disappear.

The TAZ is an encampment of guerilla ontologists: strike and run away. Keep moving the entire tribe, even if it's only data in the Web. The TAZ must be capable of defense; but both the "strike" and the "defense" should, if possible, evade the violence of the State, which is no longer a meaningful violence. The strike is made at structures of control, essentially at ideas; the defense is "invisibility," a martial art, and "invulnerability"--an "occult" art within the martial arts. The "nomadic war machine" conquers without being noticed and moves on before the map can be adjusted. As to the future--Only the autonomous can plan autonomy, organize for it, create it. It's a bootstrap operation.

The first step is somewhat akin to satori--the realization that the TAZ begins with a simple act of realization.

(Note: See Appendix C, quote by Renzo Novatore)

The Psychotopology of Everyday Life

THE CONCEPT OF THE TAZ arises first out of a critique of Revolution, and an appreciation of the Insurrection. The former labels the latter a failure; but for us uprising represents a far more interesting possibility, from the standard of a psychology of liberation, than all the "successful" revolutions of bourgeoisie, communists, fascists, etc.

The second generating force behind the TAZ springs from the historical development I call "the closure of the map." The last bit of Earth unclaimed by any nation-state was eaten up in 1899. Ours is the first century without terra incognita, without a frontier. Nationality is the highest principle of world governance--not one speck of rock in the South Seas can be left open, not one remote valley, not even the Moon and planets. This is the apotheosis of "territorial gangsterism." Not one square inch of Earth goes unpoliced or untaxed...in theory.

The "map" is a political abstract grid, a gigantic con enforced by the carrot/stick conditioning of the "Expert" State, until for most of us the map becomes the territory--no longer "Turtle Island," but "the USA." And yet because the map is an abstraction it cannot cover Earth with 1:1 accuracy. Within the fractal complexities of actual geography the map can see only dimensional grids. Hidden enfolded immensities escape the measuring rod. The map is not accurate; the map cannot be accurate.

So--Revolution is closed, but insurgency is open. For the time being we concentrate our force on temporary "power surges," avoiding all entanglements with "permanent solutions."

And--the map is closed, but the autonomous zone is open. Metaphorically it unfolds within the fractal dimensions invisible to the cartography of Control. And here we should introduce the concept of psychotopology (and -topography) as an alternative "science" to that of the State's surveying and mapmaking and "psychic imperialism." Only psychotopography can draw 1:1 maps of reality because only the human mind provides sufficient complexity to model the real. But a 1:1 map cannot "control" its territory because it is virtually identical with its territory. It can only be used to suggest, in a sense gesture towards, certain features. We are looking for "spaces" (geo-graphic, social, cultural, imaginal) with potential to flower as autonomous zones--and we are looking for times in which these spaces are relatively open, either through neglect on the part of the State or because they have somehow escaped notice by the mapmakers, or for whatever reason. Psychotopology is the art of dowsing for potential TAZs.

The closures of Revolution and of the map, however, are only the negative sources of the TAZ; much remains to be said of its positive inspirations. Reaction alone cannot provide the energy needed to "manifest" a TAZ. An uprising must be for something as well.

1. First, we can speak of a natural anthropology of the TAZ. The nuclear family is the base unit of consensus society, but not of the TAZ. ("Families!--how I hate them! the misers of love!"--Gide) The nuclear family, with its attendant "oedipal miseries," appears to have been a Neolithic invention, a response to the "agricultural revolution" with its imposed scarcity and its

imposed hierarchy. The Paleolithic model is at once more primal and more radical: the band. The typical hunter/gatherer nomadic or semi-nomadic band consists of about 50 people. Within larger tribal societies the band-structure is fulfilled by clans within the tribe, or by sodalities such as initiatic or secret societies, hunt or war societies, gender societies, "children's republics," and so on. If the nuclear family is produced by scarcity (and results in miserliness), the band is produced by abundance--and results in prodigality. The family is closed, by genetics, by the male's possession of women and children, by the hierarchic totality of agricultural/industrial society. The band is open--not to everyone, of course, but to the affinity group, the initiates sworn to a bond of love. The band is not part of a larger hierarchy, but rather part of a horizontal pattern of custom, extended kinship, contract and alliance, spiritual affinities, etc. (American Indian society preserves certain aspects of this structure even now.)

In our own post-Spectacular Society of Simulation many forces are working--largely invisibly--to phase out the nuclear family and bring back the band. Breakdowns in the structure of Work resonate in the shattered "stability" of the unit-home and unit-family. One's "band" nowadays includes friends, ex-spouses and lovers, people met at different jobs and pow-wows, affinity groups, special interest networks, mail networks, etc. The nuclear family becomes more and more obviously a trap, a cultural sinkhole, a neurotic secret implosion of split atoms--and the obvious counter-strategy emerges spontaneously in the almost unconscious rediscovery of the more archaic and yet more post-industrial possibility of the band.

2. The TAZ as festival. Stephen Pearl Andrews once offered, as an image of anarchist society, the dinner party, in which all structure of authority dissolves in conviviality and celebration (see Appendix C). Here we might also invoke Fourier and his

concept of the senses as the basis of social becoming--"touch-rut" and "gastrosophy," and his paean to the neglected implications of smell and taste. The ancient concepts of jubilee and saturnalia originate in an intuition that certain events lie outside the scope of "profane time," the measuring-rod of the State and of History. These holidays literally occupied gaps in the calendar--intercalary intervals. By the Middle Ages, nearly a third of the year was given over to holidays. Perhaps the riots against calendar reform had less to do with the "eleven lost days" than with a sense that imperial science was conspiring to close up these gaps in the calendar where the people's freedoms had accumulated--a coup d'etat, a mapping of the year, a seizure of time itself, turning the organic cosmos into a clockwork universe. The death of the festival.

Participants in insurrection invariably note its festive aspects, even in the midst of armed struggle, danger, and risk. The uprising is like a saturnalia which has slipped loose (or been forced to vanish) from its intercalary interval and is now at liberty to pop up anywhere or when. Freed of time and place, it nevertheless possesses a nose for the ripeness of events, and an affinity for the genius loci; the science of psychotopology indicates "flows of forces" and "spots of power" (to borrow occultist metaphors) which localize the TAZ spatio-temporally, or at least help to define its relation to moment and locale.

The media invite us to "come celebrate the moments of your life" with the spurious unification of commodity and spectacle, the famous non-event of pure representation. In response to this obscenity we have, on the one hand, the spectrum of refusal (chronicled by the Situationists, John Zerzan, Bob Black et al.)--and on the other hand, the emergence of a festal culture removed and even hidden from the would-be managers of our leisure. "Fight for the right to party" is in fact not a parody of the radical struggle but a new manifestation of it, appropriate to an age which offers TVs and telephones as ways to "reach out and touch" other human beings, ways to "Be There!"

Pearl Andrews was right: the dinner party is already "the seed of the new society taking shape within the shell of the old" (IWW Preamble). The sixties-style "tribal gathering," the forest conclave of eco-saboteurs, the idyllic Beltane of the neo-pagans, anarchist conferences, gay faery circles...Harlem rent parties of the twenties, nightclubs, banquets, old-time libertarian picnics--we should realize that all these are already "liberated zones" of a sort, or at least potential TAZs. Whether open only to a few friends, like a dinner party, or to thousands of celebrants, like a Be-In, the party is always "open" because it is not "ordered"; it may be planned, but unless it "happens" it's a failure. The element of spontaneity is crucial.

The essence of the party: face-to-face, a group of humans synergize their efforts to realize mutual desires, whether for good food and cheer, dance, conversation, the arts of life; perhaps even for erotic pleasure, or to create a communal artwork, or to attain the very transport of bliss--in short, a "union of egoists" (as Stirner put it) in its simplest form--or else, in Kropotkin's terms, a basic biological drive to "mutual aid." (Here we should also mention Bataille's "economy of excess" and his theory of potlatch culture.)

3. Vital in shaping TAZ reality is the concept of psychic nomad-ism (or as we jokingly call it, "rootless cosmopolitanism"). Aspects of this phenomenon have been discussed by Deleuze and Guattari in Nomadology and the War Machine, by Lyotard in Driftworks and by various authors in the "Oasis" issue of Semiotext(e). We use the term "psychic nomadism" here rather than "urban nomadism," "nomadology," "driftwork," etc., simply in order to garner all these concepts into a single loose

complex, to be studied in light of the coming-into-being of the TAZ. "The death of God," in some ways a de-centering of the entire "European" project, opened a multi-perspectived post-ideological worldview able to move "rootlessly" from philosophy to tribal myth, from natural science to Taoism--able to see for the first time through eyes like some golden insect's, each facet giving a view of an entirely other world.

But this vision was attained at the expense of inhabiting an epoch where speed and "commodity fetishism" have created a tyrannical false unity which tends to blur all cultural diversity and individuality, so that "one place is as good as another." This paradox creates "gypsies," psychic travellers driven by desire or curiosity, wanderers with shallow loyalties (in fact disloyal to the "European Project" which has lost all its charm and vitality), not tied down to any particular time and place, in search of diversity and adventure...This description covers not only the X-class artists and intellectuals but also migrant laborers, refugees, the "homeless," tourists, the RV and mobile-home culture--also people who "travel" via the Net, but may never leave their own rooms (or those like Thoreau who "have travelled much--in Concord"); and finally it includes "every-body," all of us, living through our automobiles, our vacations, our TVs, books, movies, telephones, changing jobs, changing "lifestyles," religions, diets, etc., etc.

Psychic nomadism as a tactic, what Deleuze & Guattari metaphorically call "the war machine," shifts the paradox from a passive to an active and perhaps even "violent" mode. "God"'s last throes and deathbed rattles have been going on for such a long time--in the form of Capitalism, Fascism, and Communism, for example--that there's still a lot of "creative destruction" to be carried out by post-Bakuninist post-Nietzschean commandos or apaches (literally "enemies") of the old Consensus. These nomads practice the razzia, they are corsairs, they are viruses; they have both need and desire for TAZs, camps of black tents under the desert stars, interzones, hidden fortified oases along secret caravan routes, "liberated" bits of jungle and bad-land, no-go areas, black markets, and underground bazaars.

These nomads chart their courses by strange stars, which might be luminous clusters of data in cyberspace, or perhaps hallucinations. Lay down a map of the land; over that, set a map of political change; over that, a map of the Net, especially the counter-Net with its emphasis on clandestine information-flow and logistics--and finally, over all, the 1:1 map of the creative imagination, aesthetics, values. The resultant grid comes to life, animated by unexpected eddies and surges of energy, coagulations of light, secret tunnels, surprises.

The Net and the Web

THE NEXT FACTOR CONTRIBUTING to the TAZ is so vast and ambiguous that it needs a section unto itself.

We've spoken of the Net, which can be defined as the totality of all information and communication transfer. Some of these transfers are privileged and limited to various elites, which gives the Net a hierarchic aspect. Other transactions are open to all-- so the Net has a horizontal or non-hierarchic aspect as well. Military and Intelligence data are restricted, as are banking and currency information and the like. But for the most part the telephone, the postal system, public data banks, etc. are accessible to everyone and anyone. Thus within the Net there has begun to emerge a shadowy sort of counter-Net, which we will call the Web (as if the Net were a fishing-net and the Web were spider-webs woven through the interstices and broken sections of the Net). Generally we'll use the term Web to refer to the alternate horizontal open structure of info-exchange, the non-hierarchic network, and reserve the term counter-Net to indicate clandestine illegal and rebellious use of the Web, including actual data-piracy and other forms of leeching off the Net itself. Net, Web, and counter-Net are all parts of the same whole pattern-complex--they blur into each other at innumerable points. The terms are not meant to define areas but to suggest tendencies.

(Digression: Before you condemn the Web or counter-Net for its "parasitism," which can never be a truly revolutionary force, ask yourself what "production" consists of in the Age of Simulation. What is the "productive class"? Perhaps you'll be forced to admit that these terms seem to have lost their meaning. In any case the answers to such questions are so complex that the TAZ tends to ignore them altogether and simply picks up what it can use. "Culture is our Nature"-- and we are the thieving magpies, or the hunter/gatherers of the world of CommTech.)

The present forms of the unofficial Web are, one must suppose, still rather primitive: the marginal zine network, the BBS networks, pirated software, hacking, phone-phreaking, some influence in print and radio, almost none in the other big media--no TV stations, no satellites, no fiber-optics, no cable, etc., etc. However the Net itself presents a pattern of changing/evolving relations between subjects ("users") and objects ("data"). The nature of these relations has been exhaustively explored, from McLuhan to Virilio. It would take pages and pages to "prove" what by now "everyone knows." Rather than rehash it all, I am interested in asking how these evolving relations suggest modes of implementation for the TAZ.

The TAZ has a temporary but actual location in time and a temporary but actual location in space. But clearly it must also have "location" in the Web, and this location is of a different sort, not actual but virtual, not immediate but instantaneous. The Web not only provides logistical support for the TAZ, it also helps to bring it into being; crudely speaking one might say that the TAZ "exists" in information-space as well as in the "real world." The Web can compact a great deal of time, as data, into an infinitesimal "space." We have noted that the TAZ, because it is temporary, must necessarily lack some of the advantages of a freedom which experiences duration and a more-or-less fixed locale. But the Web can provide a kind of substitute for some of this duration and locale--it can inform the TAZ, from its inception, with vast amounts of compacted time and space which have been "subtilized" as data.

At this moment in the evolution of the Web, and considering our demands for the "face-to-face" and the sensual, we must consider the Web primarily as a support system, capable of carrying information from one TAZ to another, of defending the TAZ, rendering it "invisible" or giving it teeth, as the situation might demand. But more than that: If the TAZ is a nomad camp, then the Web helps provide the epics, songs, genealogies and legends of the tribe; it provides the secret caravan routes and raiding trails which make up the flowlines of tribal economy; it even contains some of the very roads they will follow, some of the very dreams they will experience as signs and portents.

The Web does not depend for its existence on any computer technology. Word-of-mouth, mail, the marginal zine network, "phone trees," and the like already suffice to construct an information webwork. The key is not the brand or level of tech involved, but the openness and horizontality of the structure. Nevertheless, the whole concept of the Net implies the use of computers. In the SciFi imagination the Net is headed for the condition of Cyberspace (as in Tron or Neuromancer) and the pseudo-telepathy of "virtual reality." As a Cyberpunk fan I can't help but envision "reality hacking" playing a major role in the creation of TAZs. Like Gibson and Sterling I am assuming that the official Net will never succeed in shutting down the Web or the counter-Net--that data-piracy, unauthorized transmissions and the free flow of information can never be frozen. (In fact, as I understand it, chaos theory predicts that any universal Control-system is impossible.)

However, leaving aside all mere speculation about the future, we must face a very serious question about the Web and the tech it involves. The TAZ desires above all to avoid mediation, to experience its existence as immediate. The very essence of the affair is "breast-to-breast" as the sufis say, or face-to-face. But, BUT: the very essence of the Web is mediation. Machines here are our ambassadors--the flesh is irrelevant except as a terminal, with all the sinister connotations of the term.

The TAZ may perhaps best find its own space by wrapping its head around two seemingly contradictory attitudes toward Hi-Tech and its apotheosis the Net: (1) what we might call the Fifth Estate/Neo-Paleolithic Post-Situ Ultra-Green position, which construes itself as a luddite argument against mediation and against the Net; and (2) the Cyberpunk utopianists, futuro-libertarians, Reality Hackers and their allies who see the Net as a step forward in evolution, and who assume that any possible ill effects of mediation can be overcome--at least, once we've liberated the means of production.

The TAZ agrees with the hackers because it wants to come into being--in part--through the Net, even through the mediation of the Net. But it also agrees with the greens because it retains intense awareness of itself as body and feels only revulsion for CyberGnosis, the attempt to transcend the body through instantaneity and simulation. The TAZ tends to view the Tech/anti-Tech dichotomy as misleading, like most dichotomies, in which apparent opposites turn out to be falsifications or even hallucinations caused by semantics. This is a way of saying that the TAZ wants to live in this world, not in the idea of another world, some visionary world born of false unification (all green OR all metal) which can only be more pie in the sky by-&-by (or as Alice put it, "Jam yesterday or jam tomorrow, but never jam today").

The TAZ is "utopian" in the sense that it envisions an intensification of everyday life, or as the Surrealists might have said, life's penetration by the Marvelous. But it cannot be utopian in the actual meaning of the word, nowhere, or NoPlace Place. The TAZ is somewhere. It lies at the intersection of many forces, like some pagan power-spot at the junction of mysterious ley-lines, visible to the adept in seemingly unrelated bits of terrain, landscape, flows of air, water, animals. But now the lines are not all etched in time and space. Some of them exist only "within" the Web, even though they also intersect with real times and places. Perhaps some of the lines are "non-ordinary" in the sense that no convention for quantifying them exists. These lines might better be studied in the light of chaos science than of sociology, statistics, economics, etc. The patterns of force which bring the TAZ into being have something in common with those chaotic "Strange Attractors" which exist, so to speak, between the dimensions.

The TAZ by its very nature seizes every available means to realize itself--it will come to life whether in a cave or an L-5 Space City--but above all it will live, now, or as soon as possible, in however suspect or ramshackle a form, spontaneously, without regard for ideology or even anti-ideology. It will use the computer because the computer exists, but it will also use powers which are so completely unrelated to alienation or simulation that they guarantee a certain psychic paleolithism to the TAZ, a primordial-shamanic spirit which will "infect" even the Net itself (the true meaning of Cyberpunk as I read it). Because the TAZ is an intensification, a surplus, an excess, a potlatch, life spending itself in living rather than merely surviving (that snivelling shibboleth of the eighties), it cannot be defined either by Tech or anti-Tech. It contradicts itself like a true despiser of hobgoblins, because it wills itself to be, at any cost in damage to "perfection," to the immobility of the final.

In the Mandelbrot Set and its computer-graphic realization we watch--in a fractal universe--maps which are embedded and in fact hidden within maps within maps etc. to the limits of computational power. What is it for, this map which in a sense bears a 1:1 relation with a fractal dimension? What can one do with it, other than admire its psychedelic elegance?

If we were to imagine an information map--a cartographic projection of the Net in its entirety--we would have to include in it the features of chaos, which have already begun to appear, for example, in the operations of complex parallel processing, telecommunications, transfers of electronic "money," viruses, guerilla hacking and so on.

Each of these "areas" of chaos could be represented by topographs similar to the Mandelbrot Set, such that the "peninsulas" are embedded or hidden within the map--such that they seem to "disappear." This "writing"--parts of which vanish, parts of which efface themselves--represents the very process by which the Net is already compromised, incomplete to its own view, ultimately un-Controllable. In other words, the M Set, or something like it, might prove to be useful in "plotting" (in all senses of the word) the emergence of the counterNet as a chaotic process, a "creative evolution" in Prigogine's term. If nothing else the M Set serves as a metaphor for a "mapping" of the TAZ's interface with the Net as a disappearance of information. Every "catastrophe" in the Net is a node of power for the Web, the counter-Net. The Net will be damaged by chaos, while the Web may thrive on it.

Whether through simple data-piracy, or else by a more complex development of actual rapport with chaos, the Web-hacker, the cybernetician of the TAZ, will find ways to take advantage of perturbations, crashes, and breakdowns in the Net (ways to make information out of "entropy"). As a bricoleur, a scavenger of information shards, smuggler, blackmailer, perhaps even cyberterrorist, the TAZ-hacker will work for the evolution of clandestine fractal connections. These connections, and the different information that flows among and between them, will form "power outlets" for the coming-into-being of the TAZ itself--as if one were to steal electricity from the energy-monopoly to light an abandoned house for squatters.

Thus the Web, in order to produce situations conducive to the TAZ, will parasitize the Net--but we can also conceive of this strategy as an attempt to build toward the construction of an alternative and autonomous Net, "free" and no longer parasitic, which will serve as the basis for a "new society emerging from the shell of the old." The counter-Net and the TAZ can be considered, practically speaking, as ends in themselves--but theoretically they can also be viewed as forms of struggle toward a different reality.

Having said this we must still admit to some qualms about computers, some still unanswered questions, especially about the Personal Computer.

The story of computer networks, BBSs and various other experiments in electro-democracy has so far been one of hobbyism for the most part. Many anarchists and libertarians have deep faith in the PC as a weapon of liberation and self-liberation--but no real gains to show, no palpable liberty.

I have little interest in some hypothetical emergent entrepreneurial class of self-employed data/word processors who will soon be able to carry on a vast cottage industry or piecemeal shitwork for various corporations and bureaucracies. Moreover it takes no ESP to foresee that this "class" will develop its underclass--a sort of lumpen yuppetariat: housewives, for example, who will provide their families with "second incomes" by turning their own homes into electro-sweatshops, little Work-tyrannies where the "boss" is a computer network.

Also I am not impressed by the sort of information and services proffered by contemporary "radical" networks. Somewhere--one is told--there exists an "information economy." Maybe so; but the info being traded over the "alternative" BBSs seems to consist entirely of chitchat and techie-talk. Is this an economy? or merely a pastime for enthusiasts? OK, PCs have created yet another "print revolution"--OK, marginal webworks are evolving--OK, I can now carry on six phone conversations at once. But what difference has this made in my ordinary life?

Frankly, I already had plenty of data to enrich my perceptions, what with books, movies, TV, theater, telephones, the U.S. Postal Service, altered states of consciousness, and so on. Do I really need a PC in order to obtain yet more such data? You offer me secret information? Well...perhaps I'm tempted--but still I demand marvelous secrets, not just unlisted telephone numbers or the trivia of cops and politicians. Most of all I want computers to provide me with information linked to real goods--"the good things in life," as the IWW Preamble puts it. And here, since I'm accusing the hackers and BBSers of irritating intellectual vagueness, I must myself descend from the baroque clouds of Theory & Critique and explain what I mean by "real goods."

Let's say that for both political and personal reasons I desire good food, better than I can obtain from Capitalism--unpolluted food still blessed with strong and natural flavors. To complicate the game imagine that the food I crave is illegal--raw milk perhaps, or the exquisite Cuban fruit mamey, which cannot be imported fresh into the U.S. because its seed is hallucinogenic (or so I'm told). I am not a farmer. Let's pretend I'm an importer of rare perfumes and aphrodisiacs, and sharpen the play by assuming most of my stock is also illegal. Or maybe I only want to trade word processing services for organic turnips, but refuse to report the transaction to the IRS (as required by law, believe it or not). Or maybe I want to meet other humans for consensual but illegal acts of mutual pleasure (this has actually been tried, but all the hard-sex BBSs have been busted--and what use is an underground with lousy security?). In short, assume that I'm fed up with mere information, the ghost in the machine. According to you, computers should already be quite capable of facilitating my desires for food, drugs, sex, tax evasion. So what's the matter? Why isn't it happening?

The TAZ has occurred, is occurring, and will occur with or without the computer. But for the TAZ to reach its full potential it must become less a matter of spontaneous combustion and more a matter of "islands in the Net." The Net, or rather the counter-Net, assumes the promise of an integral aspect of the TAZ, an addition that will multiply its potential, a "quantum jump" (odd how this expression has come to mean a big leap) in complexity and significance. The TAZ must now exist within a world of pure space, the world of the senses. Liminal, even evanescent, the TAZ must combine information and desire in order to fulfill its adventure (its "happening"), in order to fill itself to the borders of its destiny, to saturate itself with its own becoming.

Perhaps the Neo-Paleolithic School are correct when they assert that all forms of alienation and mediation must be destroyed or abandoned before our goals can be realized--or perhaps true anarchy will be realized only in Outer Space, as some futuro-libertarians assert. But the TAZ does not concern itself very much with "was" or "will be." The TAZ is interested in results, successful raids on consensus reality, breakthroughs into more intense and more abundant life. If the computer cannot be used in this project, then the computer will have to be overcome. My intuition however suggests that the counter-Net is already coming into being, perhaps already exists--but I cannot prove it. I've based the theory of the TAZ in large part on this intuition. Of course the Web also involves non-computerized networks of exchange such as samizdat, the black market, etc.--but the full potential of non-hierarchic information networking logically leads to the computer as the tool par excellence. Now I'm waiting for the hackers to prove I'm right, that my intuition is valid. Where are my turnips?

"Gone to Croatan"

WE HAVE NO DESIRE to define the TAZ or to elaborate dogmas about how it must be created. Our contention is rather that it has been created, will be created, and is being created. Therefore it would prove more valuable and interesting to look at some TAZs past and present, and to speculate about future manifestations; by evoking a few prototypes we may be able to gauge the potential scope of the complex, and perhaps even get a glimpse of an "archetype." Rather than attempt any sort of encyclopaedism we'll adopt a scatter-shot technique, a mosaic of glimpses, beginning quite arbitrarily with the 16th-17th centuries and the settlement of the New World.

The opening of the "new" world was conceived from the start as an occultist operation. The magus John Dee, spiritual advisor to Elizabeth I, seems to have invented the concept of "magical imperialism" and infected an entire generation with it. Halkyut and Raleigh fell under his spell, and Raleigh used his connections with the "School of Night"--a cabal of advanced thinkers, aristocrats, and adepts--to further the causes of exploration, colonization and mapmaking. The Tempest was a propaganda-piece for the new ideology, and the Roanoke Colony was its first showcase experiment.

The alchemical view of the New World associated it with materia prima or hyle, the "state of Nature," innocence and all-possibility ("Virginia"), a chaos or inchoateness which the adept would transmute into "gold," that is, into spiritual perfection as well as material abundance. But this alchemical vision is also informed in part by an actual fascination with the inchoate, a sneaking sympathy for it, a feeling of yearning for its formless form which took the symbol of the "Indian" for its focus: "Man" in the state of nature, uncorrupted by "government." Caliban, the Wild Man, is lodged like a virus in the very machine of Occult Imperialism; the forest/animal/humans are invested from the very start with the magic power of the marginal, despised and outcaste. On the one hand Caliban is ugly, and Nature a "howling wilderness"--on the other, Caliban is noble and unchained, and Nature an Eden. This split in European consciousness predates the Romantic/Classical dichotomy; it's rooted in Renaissance High Magic. The discovery of America (Eldorado, the Fountain of Youth) crystallized it; and it precipitated in actual schemes for colonization.

We were taught in elementary school that the first settlements in Roanoke failed; the colonists disappeared, leaving behind them only the cryptic message "Gone To Croatan." Later reports of "grey-eyed Indians" were dismissed as legend. What really happened, the textbook implied, was that the Indians massa- cred the defenseless settlers. However, "Croatan" was not some Eldorado; it was the name of a neighboring tribe of friendly Indians. Apparently the settlement was simply moved back from the coast into the Great Dismal Swamp and absorbed into the tribe. And the grey-eyed Indians were real--they're still there, and they still call themselves Croatans.

So--the very first colony in the New World chose to renounce its contract with Prospero (Dee/Raleigh/Empire) and go over to the Wild Men with Caliban. They dropped out. They became "Indians," "went native," opted for chaos over the appalling miseries of serfing for the plutocrats and intellectuals of London.

As America came into being where once there had been "Turtle Island," Croatan remained embedded in its collective psyche. Out beyond the frontier, the state of Nature (i.e. no State) still prevailed--and within the consciousness of the settlers the option of wildness always lurked, the temptation to give up on Church, farmwork, literacy, taxes--all the burdens of civiliza- tion--and "go to Croatan" in some way or another. Moreover, as the Revolution in England was betrayed, first by Cromwell and then by Restoration, waves of Protestant radicals fled or were transported to the New World (which had now become a prison, a place of exile). Antinomians, Familists, rogue Quakers, Levellers, Diggers, and Ranters were now introduced to the occult shadow of wildness, and rushed to embrace it.

Anne Hutchinson and her friends were only the best known (i.e. the most upper-class) of the Antinomians--having had the bad luck to be caught up in Bay Colony politics--but a much more radical wing of the movement clearly existed. The incidents Hawthorne relates in "The Maypole of Merry Mount" are thoroughly historical; apparently the extremists had decided to renounce Christianity altogether and revert to paganism. If they had succeeded in uniting with their Indian allies the result might have been an Antinomian/Celtic/Algonquin syncretic religion, a sort of 17th century North American Santeria.

Sectarians were able to thrive better under the looser and more corrupt administrations in the Caribbean, where rival European interests had left many islands deserted or even unclaimed. Barbados and Jamaica in particular must have been settled by many extremists, and I believe that Levellerish and Ranterish influences contributed to the Buccaneer "utopia" on Tortuga. Here for the first time, thanks to Esquemelin, we can study a successful New World proto-TAZ in some depth. Fleeing from hideous "benefits" of Imperialism such as slavery, serfdom, racism and intolerance, from the tortures of impressment and the living death of the plantations, the Buccaneers adopted Indian ways, intermarried with Caribs, accepted blacks and Spaniards as equals, rejected all nationality, elected their captains democratically, and reverted to the "state of Nature." Having declared themselves "at war with all the world," they sailed forth to plunder under mutual contracts called "Articles" which were so egalitarian that every member received a full share and the Captain usually only 1 1/4 or 1 1/2 shares. Flogging and punishments were forbidden--quarrels were settled by vote or by the code duello.

It is simply wrong to brand the pirates as mere sea-going highwaymen or even proto-capitalists, as some historians have done. In a sense they were "social bandits," although their base communities were not traditional peasant societies but "utopias" created almost ex nihilo in terra incognita, enclaves of total liberty occupying empty spaces on the map. After the fall of Tortuga, the Buccaneer ideal remained alive all through the "Golden Age" of Piracy (ca. 1660-1720), and resulted in land-settlements in Belize, for example, which was founded by Buccaneers. Then, as the scene shifted to Madagascar--an island still unclaimed by any imperial power and ruled only by a patchwork of native kings (chiefs) eager for pirate allies--the Pirate Utopia reached its highest form.

Defoe's account of Captain Mission and the founding of Libertatia may be, as some historians claim, a literary hoax meant to propagandize for radical Whig theory--but it was embedded in The General History of the Pyrates (1724-28), most of which is still accepted as true and accurate. Moreover the story of Capt. Mission was not criticized when the book appeared and many old Madagascar hands still survived. They seem to have believed it, no doubt because they had expe-rienced pirate enclaves very much like Libertatia. Once again, rescued slaves, natives, and even traditional enemies such as the Portuguese were all invited to join as equals. (Liberating slave ships was a major preoccupation.) Land was held in common, representatives elected for short terms, booty shared; doctrines of liberty were preached far more radical than even those of Common Sense.

Libertatia hoped to endure, and Mission died in its defense. But most of the pirate utopias were meant to be temporary; in fact the corsairs' true "republics" were their ships, which sailed under Articles. The shore enclaves usually had no law at all. The last classic example, Nassau in the Bahamas, a beachfront resort of shacks and tents devoted to wine, women (and probably boys too, to judge by Birge's Sodomy and Piracy), song (the pirates were inordinately fond of music and used to hire on bands for entire cruises), and wretched excess, vanished overnight when the British fleet appeared in the Bay. Blackbeard and "Calico Jack" Rackham and his crew of pirate women moved on to wilder shores and nastier fates, while others meekly accepted the Pardon and reformed. But the Buccaneer tradition lasted, both in Madagascar where the mixed-blood children of the pirates began to carve out kingdoms of their own, and in the Caribbean, where escaped slaves as well as mixed black/white/red groups were able to thrive in the mountains and backlands as "Maroons." The Maroon community in Jamaica still retained a degree of autonomy and many of the old folkways when Zora Neale Hurston visited there in the 1920's (see Tell My Horse). The Maroons of Suriname still practice African "paganism."

Throughout the 18th century, North America also produced a number of drop-out "tri-racial isolate communities." (This clinical-sounding term was invented by the Eugenics Movement, which produced the first scientific studies of these communities. Unfortunately the "science" merely served as an excuse for hatred of racial "mongrels" and the poor, and the "solution to the problem" was usually forced sterilization.) The nuclei invariably consisted of runaway slaves and serfs, "criminals" (i.e. the very poor), "prostitutes" (i.e. white women who married non-whites), and members of various native tribes. In some cases, such as the Seminole and Cherokee, the traditional tribal structure absorbed the newcomers; in other cases, new tribes were formed. Thus we have the Maroons of the Great Dismal Swamp, who persisted through the 18th and 19th centuries, adopting runaway slaves, functioning as a way station on the Underground Railway, and serving as a religious and ideological center for slave rebellions. The religion was HooDoo, a mixture of African, native, and Christian elements, and according to the historian H. Leaming-Bey the elders of the faith and the leaders of the Great Dismal Maroons were known as "the Seven Finger High Glister."

The Ramapaughs of northern New Jersey (incorrectly known as the "Jackson Whites") present another romantic and archetypal genealogy: freed slaves of the Dutch poltroons, various Delaware and Algonquin clans, the usual "prostitutes," the "Hessians" (a catch-phrase for lost British mercenaries, drop-out Loyalists, etc.), and local bands of social bandits such as Claudius Smith's.

An African-Islamic origin is claimed by some of the groups, such as the Moors of Delaware and the Ben Ishmaels, who migrated from Kentucky to Ohio in the mid-18th century. The Ishmaels practiced polygamy, never drank alcohol, made their living as minstrels, intermarried with Indians and adopted their customs, and were so devoted to nomadism that they built their houses on wheels. Their annual migration triangulated on frontier towns with names like Mecca and Medina. In the 19th century some of them espoused anarchist ideals, and they were targeted by the Eugenicists for a particularly vicious pogrom of salvation-by-extermination. Some of the earliest Eugenics laws were passed in their honor. As a tribe they "disappeared" in the 1920's, but probably swelled the ranks of early "Black Islamic" sects such as the Moorish Science Temple. I myself grew up on legends of the "Kallikaks" of the nearby New Jersey Pine Barrens (and of course on Lovecraft, a rabid racist who was fascinated by the isolate communities). The legends turned out to be folk-memories of the slanders of the Eugenicists, whose U.S. headquarters were in Vineland, NJ, and who undertook the usual "reforms" against "miscegenation" and "feebleminded-ness" in the Barrens (including the publication of photographs of the Kallikaks, crudely and obviously retouched to make them look like monsters of misbreeding).

The "isolate communities"--at least, those which have retained their identity into the 20th century--consistently refuse to be absorbed into either mainstream culture or the black "subculture" into which modern sociologists prefer to categorize them. In the 1970's, inspired by the Native American renaissance, a number of groups--including the Moors and the Ramapaughs--applied to the B.I.A. for recognition as Indian tribes. They received support from native activists but were refused official status. If they'd won, after all, it might have set a dangerous precedent for drop-outs of all sorts, from "white Peyotists" and hippies to black nationalists, aryans, anarchists and libertarians--a "reservation" for anyone and everyone! The "European Project" cannot recognize the existence of the Wild Man--green chaos is still too much of a threat to the imperial dream of order.

Essentially the Moors and Ramapaughs rejected the "diachronic" or historical explanation of their origins in favor of a "synchronic" self-identity based on a "myth" of Indian adoption. Or to put it another way, they named themselves "Indians." If everyone who wished "to be an Indian" could accomplish this by an act of self-naming, imagine what a departure to Croatan would take place. That old occult shadow still haunts the remnants of our forests (which, by the way, have greatly increased in the Northeast since the 18-19th century as vast tracts of farmland return to scrub. Thoreau on his deathbed dreamed of the return of "...Indians...forests...": the return of the repressed).

The Moors and Ramapaughs of course have good materialist reasons to think of themselves as Indians--after all, they have Indian ancestors--but if we view their self-naming in "mythic" as well as historical terms we'll learn more of relevance to our quest for the TAZ. Within tribal societies there exist what some anthropologists call mannenbunden: totemic societies devoted to an identity with "Nature" in the act of shapeshifting, of becoming the totem-animal (werewolves, jaguar shamans, leopard men, cat-witches, etc.). In the context of an entire colonial society (as Taussig points out in Shamanism, Colonialism and the Wild Man) the shapeshifting power is seen as inhering in the native culture as a whole--thus the most repressed sector of the society acquires a paradoxical power through the myth of its occult knowledge, which is feared and desired by the colonist. Of course the natives really do have certain occult knowledge; but in response to Imperial perception of native culture as a kind of "spiritual wild(er)ness," the natives come to see themselves more and more consciously in that role. Even as they are marginalized, the Margin takes on an aura of magic. Before the whiteman, they were simply tribes of people--now, they are "guardians of Nature," inhabitants of the "state of Nature." Finally the colonist himself is seduced by this "myth." Whenever an American wants to drop out or back into Nature, invariably he "becomes an Indian." The Massachusetts radical democrats (spiritual descendents of the radical Protestants) who organized the Tea Party, and who literally believed that governments could be abolished (the whole Berkshire region declared itself in a "state of Nature"!), disguised themselves as "Mohawks." Thus the colonists, who suddenly saw themselves marginalized vis--vis the motherland, adopted the role of the marginalized natives, thereby (in a sense) seeking to participate in their occult power, their mythic radiance. From the Mountain Men to the Boy Scouts, the dream of "becoming an Indian" flows beneath myriad strands of American history, culture and consciousness.

The sexual imagery connected to "tri-racial" groups also bears out this hypothesis. "Natives" of course are always immoral, but racial renegades and drop-outs must be downright polymorphous-perverse. The Buccaneers were buggers, the Maroons and Mountain Men were miscegenists, the "Jukes and Kallikaks" indulged in fornication and incest (leading to mutations such as polydactyly), the children ran around naked and masturbated openly, etc., etc. Reverting to a "state of Nature" paradoxically seems to allow for the practice of every "unnatural" act; or so it would appear if we believe the Puritans and Eugenicists. And since many people in repressed moralistic racist societies secretly desire exactly these licentious acts, they project them outwards onto the marginalized, and thereby convince themselves that they themselves remain civilized and pure. And in fact some marginalized communities do really reject consensus morality--the pirates certainly did!--and no doubt actually act out some of civilization's repressed desires. (Wouldn't you?) Becoming "wild" is always an erotic act, an act of nakedness.

Before leaving the subject of the "tri-racial isolates," I'd like to recall Nietzsche's enthusiasm for "race mixing." Impressed by the vigor and beauty of hybrid cultures, he offered miscegenation not only as a solution to the problem of race but also as the principle for a new humanity freed of ethnic and national chauvinism--a precursor to the "psychic nomad," perhaps. Nietzsche's dream still seems as remote now as it did to him. Chauvinism still rules OK. Mixed cultures remain submerged. But the autonomous zones of the Buccaneers and Maroons, Ishmaels and Moors, Ramapaughs and "Kallikaks" remain, or their stories remain, as indications of what Nietzsche might have called "the Will to Power as Disappearance." We must return to this theme.

Music as an Organizational Principle

MEANWHILE, HOWEVER, WE TURN to the history of classical anarchism in the light of the TAZ concept.

Before the "closure of the map," a good deal of anti-authoritarian energy went into "escapist" communes such as Modern Times, the various Phalansteries, and so on. Interestingly, some of them were not intended to last "forever," but only as long as the project proved fulfilling. By Socialist/Utopian standards these experiments were "failures," and therefore we know little about them.

When escape beyond the frontier proved impossible, the era of revolutionary urban Communes began in Europe. The Communes of Paris, Lyons and Marseilles did not survive long enough to take on any characteristics of permanence, and one wonders if they were meant to. From our point of view the chief matter of fascination is the spirit of the Communes. During and after these years anarchists took up the practice of revolutionary nomadism, drifting from uprising to uprising, looking to keep alive in themselves the intensity of spirit they experienced in the moment of insurrection. In fact, certain anarchists of the Stirnerite/Nietzschean strain came to look on this activity as an end in itself, a way of always occupying an autonomous zone, the interzone which opens up in the midst or wake of war and revolution (cf. Pynchon's "zone" in Gravity's Rainbow). They declared that if any socialist revolution succeeded, they'd be the first to turn against it. Short of universal anarchy they had no intention of ever stopping. In Russia in 1917 they greeted the free Soviets with joy: this was their goal. But as soon as the Bolsheviks betrayed the Revolution, the individualist anarchists were the first to go back on the warpath. After Kronstadt, of course, all anarchists condemned the "Soviet Union" (a contradiction in terms) and moved on in search of new insurrections.

Makhno's Ukraine and anarchist Spain were meant to have duration, and despite the exigencies of continual war both succeeded to a certain extent: not that they lasted a "long time," but they were successfully organized and could have persisted if not for outside aggression. Therefore, from among the experiments of the inter-War period I'll concentrate instead on the madcap Republic of Fiume, which is much less well known, and was not meant to endure. Gabriele D'Annunzio, Decadent poet, artist, musician, aesthete, womanizer, pioneer daredevil aeronautist, black magician, genius and cad, emerged from World War I as a hero with a small army at his beck and command: the "Arditi." At a loss for adventure, he decided to capture the city of Fiume from Yugoslavia and give it to Italy. After a necromantic ceremony with his mistress in a cemetery in Venice he set out to conquer Fiume, and succeeded without any trouble to speak of. But Italy turned down his generous offer; the Prime Minister called him a fool.

In a huff, D'Annunzio decided to declare independence and see how long he could get away with it. He and one of his anarchist friends wrote the Constitution, which declared music to be the central principle of the State. The Navy (made up of deserters and Milanese anarchist maritime unionists) named themselves the Uscochi, after the long-vanished pirates who once lived on local offshore islands and preyed on Venetian and Ottoman shipping. The modern Uscochi succeeded in some wild coups: several rich Italian merchant vessels suddenly gave the Republic a future: money in the coffers! Artists, bohemians, adventurers, anarchists (D'Annunzio corresponded with Malatesta), fugitives and Stateless refugees, homosexuals, military dandies (the uniform was black with pirate skull-&-crossbones--later stolen by the SS), and crank reformers of every stripe (including Buddhists, Theosophists and Vedantists) began to show up at Fiume in droves. The party never stopped. Every morning D'Annunzio read poetry and manifestos from his balcony; every evening a concert, then fireworks. This made up the entire activity of the government. Eighteen months later, when the wine and money had run out and the Italian fleet finally showed up and lobbed a few shells at the Municipal Palace, no one had the energy to resist.

D'Annunzio, like many Italian anarchists, later veered toward fascism--in fact, Mussolini (the ex-Syndicalist) himself seduced the poet along that route. By the time D'Annunzio realized his error it was too late: he was too old and sick. But Il Duce had him killed anyway--pushed off a balcony--and turned him into a "martyr." As for Fiume, though it lacked the seriousness of the free Ukraine or Barcelona, it can probably teach us more about certain aspects of our quest. It was in some ways the last of the pirate utopias (or the only modern example)--in other ways, perhaps, it was very nearly the first modern TAZ.

I believe that if we compare Fiume with the Paris uprising of 1968 (also the Italian urban insurrections of the early seventies), as well as with the American countercultural communes and their anarcho-New Left influences, we should notice certain similarities, such as:--the importance of aesthetic theory (cf. the Situationists)--also, what might be called "pirate economics," living high off the surplus of social overproduction--even the popularity of colorful military uniforms--and the concept of music as revolutionary social change--and finally their shared air of impermanence, of being ready to move on, shape-shift, re-locate to other universities, mountaintops, ghettos, factories, safe houses, abandoned farms--or even other planes of reality. No one was trying to impose yet another Revolutionary Dictatorship, either at Fiume, Paris, or Millbrook. Either the world would change, or it wouldn't. Meanwhile keep on the move and live intensely.

The Munich Soviet (or "Council Republic") of 1919 exhibited certain features of the TAZ, even though--like most revolutions-- its stated goals were not exactly "temporary." Gustav Landauer's participation as Minister of Culture along with Silvio Gesell as Minister of Economics and other anti-authoritarian and extreme libertarian socialists such as the poet/playwrights Erich Mühsam and Ernst Toller, and Ret Marut (the novelist B. Traven), gave the Soviet a distinct anarchist flavor. Landauer, who had spent years of isolation working on his grand synthesis of Nietzsche, Proudhon, Kropotkin, Stirner, Meister Eckhardt, the radical mystics, and the Romantic volk-philosophers, knew from the start that the Soviet was doomed; he hoped only that it would last long enough to be understood. Kurt Eisner, the martyred founder of the Soviet, believed quite literally that poets and poetry should form the basis of the revolution. Plans were launched to devote a large piece of Bavaria to an experiment in anarcho-socialist economy and community. Landauer drew up proposals for a Free School system and a People's Theater. Support for the Soviet was more or less confined to the poorest working-class and bohemian neighborhoods of Munich, and to groups like the Wandervogel (the neo-Romantic youth movement), Jewish radicals (like Buber), the Expressionists, and other marginals. Thus historians dismiss it as the "Coffeehouse Republic" and belittle its significance in comparison with Marxist and Spartacist participation in Germany's post-War revolution(s). Outmaneuvered by the Communists and eventually murdered by soldiers under the influence of the occult/fascist Thule Society, Landauer deserves to be remembered as a saint. Yet even anarchists nowadays tend to misunderstand and condemn him for "selling out" to a "socialist government." If the Soviet had lasted even a year, we would weep at the mention of its beauty--but before even the first flowers of that Spring had wilted, the geist and the spirit of poetry were crushed, and we have forgotten. Imagine what it must have been to breathe the air of a city in which the Minister of Culture has just predicted that schoolchildren will soon be memorizing the works of Walt Whitman. Ah for a time machine...

The Will to Power as Disappearance

FOUCAULT, BAUDRILLARD, ET AL. have discussed various modes of "disappearance" at great length. Here I wish to suggest that the TAZ is in some sense a tactic of disappearance. When the Theorists speak of the disappearance of the Social they mean in part the impossibility of the "Social Revolution," and in part the impossibility of "the State"-- the abyss of power, the end of the discourse of power. The anarchist question in this case should then be: Why bother to confront a "power" which has lost all meaning and become sheer Simulation? Such confrontations will only result in dangerous and ugly spasms of violence by the emptyheaded shit-for-brains who've inherited the keys to all the armories and prisons. (Perhaps this is a crude american misunderstanding of sublime and subtle Franco-Germanic Theory. If so, fine; whoever said understanding was needed to make use of an idea?)

As I read it, disappearance seems to be a very logical radical option for our time, not at all a disaster or death for the radical project. Unlike the morbid deathfreak nihilistic interpretation of Theory, mine intends to mine it for useful strategies in the always-ongoing "revolution of everyday life": the struggle that cannot cease even with the last failure of political or social revolution because nothing except the end of the world can bring an end to everyday life, nor to our aspirations for the good things, for the Marvelous. And as Nietzsche said, if the world could come to an end, logically it would have done so; it has not, so it does not. And so, as one of the sufis said, no matter how many draughts of forbidden wine we drink, we will carry this raging thirst into eternity.

Zerzan and Black have independently noted certain "elements of Refusal" (Zerzan's term) which perhaps can be seen as somehow symptomatic of a radical culture of disappearance, partly unconscious but partly conscious, which influences far more people than any leftist or anarchist idea. These gestures are made against institutions, and in that sense are "negative"-- but each negative gesture also suggests a "positive" tactic to replace rather than merely refuse the despised institution.

For example, the negative gesture against schooling is "voluntary illiteracy." Since I do not share the liberal worship of literacy for the sake of social amelioration, I cannot quite share the gasps of dismay heard everywhere at this phenomenon: I sympathize with children who refuse books along with the garbage in the books. There are however positive alternatives which make use of the same energy of disappearance. Home-schooling and craft-apprenticeship, like truancy, result in an absence from the prison of school. Hacking is another form of "education" with certain features of "invisibility."

A mass-scale negative gesture against politics consists simply of not voting. "Apathy" (i.e. a healthy boredom with the weary Spectacle) keeps over half the nation from the polls; anarchism never accomplished as much! (Nor did anarchism have anything to do with the failure of the recent Census.) Again, there are positive parallels: "networking" as an alternative to politics is practiced at many levels of society, and non-hierarchic organization has attained popularity even outside the anarchist movement, simply because it works. (ACT UP and Earth First! are two examples. Alcoholics Anonymous, oddly enough, is another.)

Refusal of Work can take the forms of absenteeism, on-job drunkenness, sabotage, and sheer inattention--but it can also give rise to new modes of rebellion: more self-employment, participation in the "black" economy and "lavoro nero," welfare scams and other criminal options, pot farming, etc.--all more or less "invisible" activities compared to traditional leftist confrontational tactics such as the general strike.

Refusal of the Church? Well, the "negative gesture" here probably consists of...watching television. But the positive alternatives include all sorts of non-authoritarian forms of spirituality, from "unchurched" Christianity to neo-paganism. The "Free Religions" as I like to call them--small, self-created, half-serious/half-fun cults influenced by such currents as Discordianism and anarcho-Taoism--are to be found all over marginal America, and provide a growing "fourth way" outside the mainstream churches, the televangelical bigots, and New Age vapidity and consumerism. It might also be said that the chief refusal of orthodoxy consists of the construction of "private moralities" in the Nietzschean sense: the spirituality of "free spirits."

The negative refusal of Home is "homelessness," which most consider a form of victimization, not wishing to be forced into nomadology. But "homelessness" can in a sense be a virtue, an adventure--so it appears, at least, to the huge international movement of the squatters, our modern hobos.

The negative refusal of the Family is clearly divorce, or some other symptom of "breakdown." The positive alternative springs from the realization that life can be happier without the nuclear family, whereupon a hundred flowers bloom--from single parentage to group marriage to erotic affinity group. The "European Project" fights a major rearguard action in defense of

"Family"--oedipal misery lies at the heart of Control. Alternatives exist--but they must remain in hiding, especially since the War against Sex of the 1980's and 1990's.

What is the refusal of Art? The "negative gesture" is not to be found in the silly nihilism of an "Art Strike" or the defacing of some famous painting--it is to be seen in the almost universal glassy-eyed boredom that creeps over most people at the very mention of the word. But what would the "positive gesture" consist of? Is it possible to imagine an aesthetics that does not engage, that removes itself from History and even from the Market? or at least tends to do so? which wants to replace representation with presence? How does presence make itself felt even in (or through) representation?

"Chaos Linguistics" traces a presence which is continually disappearing from all orderings of language and meaning-systems; an elusive presence, evanescent, latif ("subtle," a term in sufi alchemy)--the Strange Attractor around which memes accrue, chaotically forming new and spontaneous orders. Here we have an aesthetics of the borderland between chaos and order, the margin, the area of "catastrophe" where the breakdown of the system can equal enlightenment. (Note: for an explanation of "Chaos Linguistics" see Appendix A, then please read this paragraph again.)

The disappearance of the artist IS "the suppression and realization of art," in Situationist terms. But from where do we vanish? And are we ever seen or heard of again? We go to Croatan--what's our fate? All our art consists of a goodbye note to history--"Gone To Croatan"--but where is it, and what will we do there?

First: We're not talking here about literally vanishing from the world and its future:--no escape backward in time to paleolithic "original leisure society"--no forever utopia, no backmountain hideaway, no island; also, no post-Revolutionary utopia--most likely no Revolution at all!--also, no VONU, no anarchist Space Stations--nor do we accept a "Baudrillardian disappearance" into the silence of an ironic hyperconformity. I have no quarrel with any Rimbauds who escape Art for whatever Abyssinia they can find. But we can't build an aesthetics, even an aesthetics of disappearance, on the simple act of never coming back. By saying we're not an avant-garde and that there is no avant-garde, we've written our "Gone To Croatan"--the question then becomes, how to envision "everyday life" in Croatan? particularly if we cannot say that Croatan exists in Time (Stone Age or Post-Revolution) or Space, either as utopia or as some forgotten midwestern town or as Abyssinia? Where and when is the world of unmediated creativity? If it can exist, it does exist--but perhaps only as a sort of alternate reality which we so far have not learned to perceive. Where would we look for the seeds--the weeds cracking through our sidewalks--from this other world into our world? the clues, the right directions for searching? a finger pointing at the moon?

I believe, or would at least like to propose, that the only solution to the "suppression and realization" of Art lies in the emergence of the TAZ. I would strongly reject the criticism that the TAZ itself is "nothing but" a work of art, although it may have some of the trappings. I do suggest that the TAZ is the only possible "time" and "place" for art to happen for the sheer pleasure of creative play, and as an actual contribution to the forces which allow the TAZ to cohere and manifest.

Art in the World of Art has become a commodity; but deeper than that lies the problem of re-presentation itself, and the refusal of all mediation. In the TAZ art as a commodity will simply become impossible; it will instead be a condition of life.

Mediation is harder to overcome, but the removal of all barriers between artists and "users" of art will tend toward a condition in which (as A.K. Coomaraswamy described it) "the artist is not a special sort of person, but every person is a special sort of artist."

In sum: disappearance is not necessarily a "catastrophe"--except in the mathematical sense of "a sudden topological change." All the positive gestures sketched here seem to involve various degrees of invisibility rather than traditional revolutionary confrontation. The "New Left" never really believed in its own existence till it saw itself on the Evening News. The New Autonomy, by contrast, will either infiltrate the media and subvert "it" from within--or else never be "seen" at all. The TAZ exists not only beyond Control but also beyond definition, beyond gazing and naming as acts of enslaving, beyond the understanding of the State, beyond the State's ability to see.

The Will to Power as Disappearance

THE TAZ AS A CONSCIOUS radical tactic will emerge under certain conditions:

1. Psychological liberation. That is, we must realize (make real) the moments and spaces in which freedom is not only possible but actual. We must know in what ways we are genuinely oppressed, and also in what ways we are self-repressed or ensnared in a fantasy in which ideas oppress us. WORK, for example, is a far more actual source of misery for most of us than legislative politics. Alienation is far more dangerous for us than toothless outdated dying ideologies. Mental addiction to "ideals"--which in fact turn out to be mere projections of our resentment and sensations of victimization--will never further our project. The TAZ is not a harbinger of some pie-in-the-sky Social Utopia to which we must sacrifice our lives that our

children's children may breathe a bit of free air. The TAZ must be the scene of our present autonomy, but it can only exist on the condition that we already know ourselves as free beings.

2. The counter-Net must expand. At present it reflects more abstraction than actuality. Zines and BBSs exchange information, which is part of the necessary groundwork of the TAZ, but very little of this information relates to concrete goods and services necessary for the autonomous life. We do not live in CyberSpace; to dream that we do is to fall into CyberGnosis, the false transcendence of the body. The TAZ is a physical place and we are either in it or not. All the senses must be involved. The Web is like a new sense in some ways, but it must be added to the others--the others must not be subtracted from it, as in some horrible parody of the mystic trance. Without the Web, the full realization of the TAZ-complex would be impossible. But the Web is not the end in itself. It's a weapon.

3. The apparatus of Control--the "State"--must (or so we must assume) continue to deliquesce and petrify simultaneously, must progress on its present course in which hysterical rigidity comes more and more to mask a vacuity, an abyss of power. As power "disappears," our will to power must be disappearance.

We've already dealt with the question of whether the TAZ can be viewed "merely" as a work of art. But you will also demand to know whether it is more than a poor rat-hole in the Babylon of Information, or rather a maze of tunnels, more and more connected, but devoted only to the economic dead-end of piratical parasitism? I'll answer that I'd rather be a rat in the wall than a rat in the cage--but I'll also insist that the TAZ transcends these categories.

A world in which the TAZ succeeded in putting down roots might resemble the world envisioned by "P.M." in his fantasy novel bolo'bolo. Perhaps the TAZ is a "proto-bolo." But inasmuch as the TAZ exists now, it stands for much more than the mundanity of negativity or countercultural drop-out-ism. We've mentioned the festal aspect of the moment which is unControlled, and which adheres in spontaneous self-ordering, however brief. It is "epiphanic"--a peak experience on the social as well as individual scale.

Liberation is realized struggle--this is the essence of Nietzsche's "self-overcoming." The present thesis might also take for a sign Nietzsche's wandering. It is the precursor of the drift, in the Situ sense of the derive and Lyotard's definition of driftwork. We can foresee a whole new geography, a kind of pilgrimage-map in which holy sites are replaced by peak experiences and TAZs: a real science of psychotopography, perhaps to be called "geo-autonomy" or "anarchomancy."

The TAZ involves a kind of ferality, a growth from tameness to wild(er)ness, a "return" which is also a step forward. It also demands a "yoga" of chaos, a project of "higher" orderings (of consciousness or simply of life) which are approached by "surfing the wave-front of chaos," of complex dynamism. The TAZ is an art of life in continual rising up, wild but gentle--a seducer not a rapist, a smuggler rather than a bloody pirate, a dancer not an eschatologist.

Let us admit that we have attended parties where for one brief night a republic of gratified desires was attained. Shall we not confess that the politics of that night have more reality and force for us than those of, say, the entire U.S. Government? Some of the "parties" we've mentioned lasted for two or three years. Is this something worth imagining, worth fighting for? Let us study invisibility, webworking, psychic nomadism--and who knows what we might attain?

--Spring Equinox, 1990

APPENDIX A. CHAOS LINGUISTICS

NOT YET A SCIENCE but a proposition: That certain problems in linguistics might be solved by viewing language as a complex dynamical system or "Chaos field."

Of all the responses to Saussure's linguistics, two have special interest here: the first, "antilinguistics," can be traced--in the modern period--from Rimbaud's departure for Abyssinia; to Nietzsche's "I fear that while we still have grammar we have not yet killed God"; to dada; to Korzybski's "the Map is not the Territory"; to Burroughs' cut-ups and "breakthrough in the Gray Room"; to Zerzan's attack on language itself as representation and mediation.

The second, Chomskyan Linguistics, with its belief in "universal grammar" and its tree diagrams, represents (I believe) an attempt to "save" language by discovering "hidden invariables," much in the same way certain scientists are trying to "save" physics from the "irrationality" of quantum mechanics. Although as an anarchist Chomsky might have been expected to side with the nihilists, in fact his beautiful theory has more in common with platonism or sufism than with anarchism. Traditional metaphysics describes language as pure light shining through the colored glass of the archetypes; Chomsky speaks of "innate" grammars. Words are leaves, branches are sentences, mother tongues are limbs, language families are trunks, and the roots are in "heaven"...or the DNA. I call this "hermetalinguistics"--hermetic and metaphysical. Nihilism (or "HeavyMetalinguistics" in honor of Burroughs) seems to me to have brought language to a dead end and threatened to render it "impossible" (a great feat, but a depressing one)--while Chomsky holds out the promise and hope of a last-minute revelation, which I find equally difficult to accept. I too would like to "save" language, but without recourse to any "Spooks," or supposed rules about God, dice, and the Universe.

Returning to Saussure, and his posthumously published notes on anagrams in Latin poetry, we find certain hints of a process which somehow escapes the sign/signifier dynamic. Saussure was confronted with the suggestion of some sort of "meta"-linguistics which happens within language rather than being imposed as a categorical imperative from "outside." As soon as language begins to play, as in the acrostic poems he examined, it seems to resonate with self-amplifying complexity. Saussure tried to quantify the anagrams but his figures kept running away from him (as if perhaps nonlinear equations were involved). Also, he began to find the anagrams everywhere, even in Latin prose. He began to wonder if he were hallucinating--or if anagrams were a natural unconscious process of parole. He abandoned the project.

I wonder: if enough of this sort of data were crunched through a computer, would we begin to be able to model language in terms of complex dynamical systems? Grammars then would not be "innate," but would emerge from chaos as spontaneously evolving "higher orders," in Prigogine's sense of "creative evolution." Grammars could be thought of as "Strange Attractors," like the hidden pattern which "caused" the anagrams--patterns which are "real" but have "existence" only in terms of the sub-patterns they manifest. If meaning is elusive, perhaps it is because consciousness itself, and therefore language, is fractal.

I find this theory more satisfyingly anarchistic than either anti-linguistics or Chomskyanism. It suggests that language can overcome representation and mediation, not because it is innate, but because it is chaos. It would suggest that all dadaistic experimentation (Feyerabend described his school of scientific epistemology as "anarchist dada") in sound poetry, gesture, cut-up, beast languages, etc.--all this was aimed neither at discovering nor destroying meaning, but at creating it. Nihilism points out gloomily that language "arbitrarily" creates meaning. Chaos Linguistics happily agrees, but adds that language can overcome language, that language can create freedom out of semantic tyranny's confusion and decay.

APPENDIX B. APPLIED HEDONICS

THE BONNOT GANG WERE vegetarians and drank only water. They came to a bad (tho' picturesque) end. Vegetables and water, in themselves excellent things--pure zen really--shouldn't be consumed as martyrdom but as an epiphany. Self-denial as radical praxis, the Leveller impulse, tastes of millenarian gloom-- and this current on the Left shares an historical wellspring with the neo-puritan fundamentalism and moralic reaction of our decade. The New Ascesis, whether practiced by anorexic health-cranks, thin-lipped police sociologists, downtown straight-edge nihilists, cornpone fascist baptists, socialist torpedoes, drug-free Republicans...in every case the motive force is the same: resentment.

In the face of contemporary pecksniffian anaesthesia we'll erect a whole gallery of forebears, heros who carried on the struggle against bad consciousness but still knew how to party, a genial gene pool, a rare and difficult category to define, great minds not just for Truth but for the truth of pleasure, serious but not sober, whose sunny disposition makes them not sluggish but sharp, brilliant but not tormented. Imagine a Nietzsche with good digestion. Not the tepid Epicureans nor the bloated Sybarites. Sort of a spiritual hedonism, an actual Path of Pleasure, vision of a good life which is both noble and possible, rooted in a sense of the magnificent over-abundance of reality.

Shaykh Abu Sa'id of Khorassan
Charles Fourier
Brillat-Savarin
Rabelais

Abu Nuwas
Aga Khan III
R. Vaneigem
Oscar Wilde
Omar Khayyam
Sir Richard Burton
Emma Goldman
add your own favorites

APPENDIX C. EXTRA QUOTES

As for us, He has appointed the job of permanent unemployment.
If he wanted us to work, after all,
He would not have created this wine.
With a skinfull of this, Sir,
would you rush out to commit economics?

--Jalaloddin Rumi, Diwan-e Shams

Here with a Loaf of Bread beneath the Bough,
A flask of Wine, A Book of Verse--and Thou
Beside me singing in the Wilderness--
And Wilderness is Paradise enow.
Ah, my Beloved, fill the cup that clears
To-day of past Regrets and future Fears--
Tomorrow?--Why, Tomorrow I may be
Myself with Yesterday's Sev'n Thousand Years.
Ah, Love! could thou and I with Fate conspire
To grasp this sorry Scheme of Things entire,
Would not we shatter it to bits--and then
Re-mould it nearer to the Heart's Desire!

--Omar FitzGerald

History, materialism, monism, positivism, and all the "isms" of this world are old and rusty tools which I don't need or mind anymore. My principle is life, my end is death. I wish to live my life intensely for to embrace my life tragically.

You are waiting for the revolution? My own began a long time ago! When you will be ready (God, what an endless wait!) I won't mind going along with you for awhile. But when you'll stop, I shall continue on my insane and triumphal way toward the great and sublime conquest of the nothing! Any society that you build will have its limits. And outside the limits of any society the unruly and heroic tramps will wander, with their wild & virgin thoughts--they who cannot live without planning ever new and dreadful outbursts of rebellion!

I shall be among them!

And after me, as before me, there will be those saying to their fellows: "So turn to yourselves rather than to your Gods or to your idols. Find what hides in yourselves; bring it to light; show yourselves!"

Because every person; who, searching his own inwardness, extracts what was mysteriously hidden therein; is a shadow eclipsing any form of society which can exist under the sun! All societies tremble when the scornful aristocracy of the tramps, the inaccessibles, the uniques, the rulers over the ideal, and the conquerors of the nothing resolutely advances.

So, come on iconoclasts, forward!

"Already the foreboding sky grows dark and silent!"

--Renzo Novatore Arcola, January, 1920

PIRATE RANT

Captain Bellamy

Daniel Defoe, writing under the pen name Captain Charles Johnson, wrote what became the first standard historical text on pirates, A General History of the Robberies and Murders of the Most Notorious Pirates. According to Patrick Pringle's Jolly Roger, pirate recruitment was most effective among the unemployed, escaped bondsmen, and transported criminals. The high seas made for an instantaneous levelling of class inequalities. Defoe relates that a pirate named Captain Bellamy made this speech to the captain of a merchant vessel he had taken as a prize. The captain of the merchant vessel had just declined an invitation to join the pirates.

I am sorry they won't let you have your sloop again, for I scorn to do any one a mischief, when it is not to my advantage; damn the sloop, we must sink her, and she might be of use to you. Though you are a sneaking puppy, and so are all those who will submit to be governed by laws which rich men have made for their own security; for the cowardly whelps have not the courage otherwise to defend what they get by knavery; but damn ye altogether: damn them for a pack of crafty rascals, and you, who serve them, for a parcel of hen-hearted numbskulls. They vilify us, the scoundrels do, when there is only this difference, they rob the poor under the cover of law, forsooth, and we plunder the rich under the protection of our own courage. Had you not better make then one of us, than sneak after these villains for employment?

When the captain replied that his conscience would not let him break the laws of God and man, the pirate Bellamy continued:

You are a devilish conscience rascal, I am a free prince, and I have as much authority to make war on the whole world, as he who has a hundred sail of ships at sea, and an army of 100,000 men in the field; and this my conscience tells me: but there is no arguing with such snivelling puppies, who allow superiors to kick them about deck at pleasure.

THE DINNER PARTY

The highest type of human society in the existing social order is found in the parlor. In the elegant and refined reunions of the aristocratic classes there is none of the impertinent interference of legislation. The Individuality of each is fully admitted. Intercourse, therefore, is perfectly free. Conversation is continuous, brilliant, and varied. Groups are formed according to attraction. They are continuously broken up, and re-formed through the operation of the same subtle and all-pervading influence. Mutual deference pervades all classes, and the most perfect harmony, ever yet attained, in complex human relations, prevails under precisely those circumstances which Legislators and Statesmen dread as the conditions of inevitable anarchy and confusion. If there are laws of etiquette at all, they are mere suggestions of principles admitted into and judged of for himself or herself, by each individual mind.

Is it conceivable that in all the future progress of humanity, with all the innumerable elements of development which the present age is unfolding, society generally, and in all its relations, will not attain as high a grade of perfection as certain portions of society, in certain special relations, have already attained?

Suppose the intercourse of the parlor to be regulated by specific legislation. Let the time which each gentleman shall be allowed to speak to each lady be fixed by law; the position in which they should sit or stand be precisely regulated; the subjects which they shall be allowed to speak of, and the tone of voice and accompanying gestures with which each may be treated, carefully defined, all under pretext of preventing disorder and encroachment upon each other's privileges and rights, then can any thing be conceived better calculated or more certain to convert social intercourse into intolerable slavery and hopeless confusion?

--S. Pearl Andrews The Science of Society

www.ingramcontent.com/pod-product-compliance
Lightning Source LLC
Chambersburg PA
CBHW051246020426